Manhattan Review®

Test Prep & Admissions Consulting

Turbocharge Your DSAT: Math Question Bank

Published on February 11th, 2025

Version 69

- ☐ *Designed as per the Digital SAT*
- ☐ *Classified into four sub-sections:*
  - *Heart of Algebra*
  - *Problem Solving & Data Analysis*
  - *Passport to Advanced Math*
  - *Advanced topics in Math*
- ☐ *Complete & Challenging Training Sets: 300 questions*
- ☐ *Ample questions on new additions:*
  - *Trigonometry & Complex Numbers*
  - *Linear graphic & Quadratic graphic equations*
  - *The concept of Higher order Thinking*
  - *Application of Functions*
- ☐ *Each question tagged with 'With calculator' or 'Without calculator'*

www.manhattanreview.com

Copyright and Terms of Use

Copyright and Trademark

Terms of Use

No Warranties

Limitation on Liability

International Standard Book Number (ISBN): 978-1-62926-118-8

Published on February 11th, 2025.

Manhattan Review, 2 Park Avenue, Suite 2010, New York, NY 10016.
Phone: +1 (212) 316-2000. E-Mail: info@manhattanreview.com. Web: www.manhattanreview.com

About the Turbocharge your SAT Series

The Turbocharge Your SAT Series was created to provide students with comprehensive and highly effective SAT preparation for maximum SAT performance. Thousands of students around the world have received substantial score improvements by using Manhattan Review's SAT prep books. Now in its updated 2nd edition for the new SAT in 2016, the full series of 11 guides is designed to provide SAT students with rigorous, thorough, and accessible SAT instruction for top SAT scores. Manhattan Review's SAT prep books precisely target each testing area and deconstruct the different test sections in a manner that is both student-centered and results-driven, teaching test-takers everything they need to know in order to significantly boost their scores. Covering all of the necessary material in mathematics and verbal skills from the most basic through the most advanced levels, the Turbocharge Your SAT Series is the top study resource for all stages of SAT preparation. Students who work through the complete series develop all of the skills, knowledge, and strategies needed for their best possible SAT scores.

- ☐ **DSAT Math Essentials (ISBN: 978-1-62926-115-7)**
- ☐ **DSAT Math Guide (ISBN: 978-1-62926-116-4)**
- ☐ **DSAT Reading & Writing Guide (ISBN: 978-1-62926-117-1)**
- ■ **DSAT Math Question Bank (ISBN: 978-1-62926-118-8)**

About the Company

Manhattan Review's origin can be traced directly back to an Ivy League MBA classroom in 1999. While teaching advanced quantitative subjects to MBAs at Columbia Business School in New York City, Professor Dr. Joern Meissner developed a reputation for explaining complicated concepts in an understandable way. Prof. Meissner's students challenged him to assist their friends, who were frustrated with conventional test preparation options. In response, Prof. Meissner created original lectures that focused on presenting standardized test content in a simplified and intelligible manner, a method vastly different from the voluminous memorization and so-called tricks commonly offered by others. The new methodology immediately proved highly popular with students, inspiring the birth of Manhattan Review.

Since its founding, Manhattan Review has grown into a multi-national educational services firm, focusing on preparation for the major undergraduate and graduate admissions tests, college admissions consulting, and application advisory services, with thousands of highly satisfied students all over the world. Our SAT instruction is continuously expanded and updated by the Manhattan Review team, an enthusiastic group of master SAT professionals and senior academics. Our team ensures that Manhattan Review offers the most time-efficient and cost-effective preparation available for the SAT. Please visit www.ManhattanReview.com for further details.

About the Founder

Professor Dr. Joern Meissner has more than 25 years of teaching experience at the graduate and undergraduate levels. He is the founder of Manhattan Review, a worldwide leader in test prep services, and he created the original lectures for its first test preparation classes. Prof. Meissner is a graduate of Columbia Business School in New York City, where he received a PhD in Management Science. He has since served on the faculties of prestigious business schools in the United Kingdom and Germany. He is a recognized authority in the areas of supply chain management, logistics, and pricing strategy. Prof. Meissner thoroughly enjoys his research, but he believes that grasping an idea is only half of the fun. Conveying knowledge to others is even more fulfilling. This philosophy was crucial to the establishment of Manhattan Review, and remains its most cherished principle.

International Phone Numbers and Official Manhattan Review Websites

Manhattan Headquarters	+1-212-316-2000	www.manhattanreview.com
USA & Canada	+1-800-246-4600	www.manhattanreview.com
Argentina	+1-212-316-2000	www.review.com.ar
Australia	+61-3-9001-6618	www.manhattanreview.com
Austria	+43-720-115-549	www.review.at
Belgium	+32-2-808-5163	www.manhattanreview.be
Brazil	+1-212-316-2000	www.manhattanreview.com.br
Chile	+1-212-316-2000	www.manhattanreview.cl
China	+86-20-2910-1913	www.manhattanreview.cn
Czech Republic	+1-212-316-2000	www.review.cz
France	+33-1-8488-4204	www.review.fr
Germany	+49-89-3803-8856	www.review.de
Greece	+1-212-316-2000	www.review.com.gr
Hong Kong	+852-5808-2704	www.review.hk
Hungary	+1-212-316-2000	www.review.co.hu
India	+1-212-316-2000	www.review.in
Indonesia	+1-212-316-2000	www.manhattanreview.id
Ireland	+1-212-316-2000	www.gmat.ie
Italy	+39-06-9338-7617	www.manhattanreview.it
Japan	+81-3-4589-5125	www.manhattanreview.jp
Malaysia	+1-212-316-2000	www.review.my
Netherlands	+31-20-808-4399	www.manhattanreview.nl
New Zealand	+1-212-316-2000	www.review.co.nz
Philippines	+1-212-316-2000	www.review.ph
Poland	+1-212-316-2000	www.review.pl
Portugal	+1-212-316-2000	www.review.pt
Qatar	+1-212-316-2000	www.review.qa
Russia	+1-212-316-2000	www.manhattanreview.ru
Singapore	+65-3158-2571	www.gmat.sg
South Africa	+1-212-316-2000	www.manhattanreview.co.za
South Korea	+1-212-316-2000	www.manhattanreview.kr
Sweden	+1-212-316-2000	www.gmat.se
Spain	+34-911-876-504	www.review.es
Switzerland	+41-435-080-991	www.review.ch
Taiwan	+1-212-316-2000	www.gmat.tw
Thailand	+66-6-0003-5529	www.manhattanreview.com
Turkey	+1-212-316-2000	www.review.com.tr
United Arab Emirates	+1-212-316-2000	www.manhattanreview.ae
United Kingdom	+44-20-7060-9800	www.manhattanreview.co.uk
Rest of World	+1-212-316-2000	www.manhattanreview.com

Contents

Chapter 1

Welcome

Dear Students,

Here at Manhattan Review, we constantly strive to provide you with the best educational content for standardized test preparation. We make a tremendous effort to keep making things better and better for you. This is especially important for an examination such as the SAT. As you know, from Spring'24 on, the SAT went through a major change. The Digital SAT is even more challenging.

Manhattan Review's DSAT-Math Question Bank has a collection of 300 questions classified into four sub-sections: **Algebra, Advanced Math, Problem Solving & Data Analysis, and Geometry & Trigonometry**.

In a nutshell, Manhattan Review's "DSAT-Math Question Bank" is holistic and comprehensive for the practice; it was created so because we listen to what students need. Should you have any queries, please feel free to write to us at *info@manhattanreview.com.*

Happy Learning!

Professor Dr. Joern Meissner
& The Manhattan Review Team

Chapter 2

Introduction to The Digital SAT

In 2023–24, the College Board changed the previous SAT to a digital format. The digital version of the SAT is reduced to 2 hours, 14 minutes, as opposed to the previously 3 hours, 15 minutes. This change is welcome because it validates the relevance and significance of standardized testing.

When Can a Student Take the Digital SAT?

The College board publishes test dates for the digital SAT, as before.

What changes did the College Board make to the digital SAT?

In addition to reducing testing time to 2 hours, 14 minutes from the previous 3 hours, 15 minutes, the digital SAT will allow more time per question.

The following are the significant changes compared to the previous SAT content:

- Shorter reading passages with a single (discrete) question per text
- Texts from a broader range of college-level exams
- Calculator access for the entire math section
- The digital SAT scores are provided more quickly, allowing students and universities to make quicker college decisions.

Advantages of digital SAT

Safe: The paper-and-pencil version of the test could be canceled if there are problems outside or inside the school. The digital SAT is safer and better protected because each student will have a unique test form and will use their own or a test center-issued device.

Faster Score: Students get their scores in days as opposed to weeks.

What are the differences between the digital SAT and the previous SAT?

The digital SAT is a simpler version of the previous SAT. The new structure will place more emphasis on college-ready abilities and less on tenacity and speed. The previous SAT Reading section, with its long passages and frequently vague history passages, demanded patience to solve. The digital SAT incorporates texts from the humanities and sciences that dive into more modern topics in finance, economics, and the environment, as well as a greater variety of short passages.

Also, there is relatively greater emphasis on using both textual and graphic content to answer questions.

Since calculators are used for the entire math section, it seems like the test has been changed in a big way. In addition to the focus on algebra and word problems on the SAT, there may be a shift to more straightforward questions, like those on the ACT, and more geometry and trigonometry questions.

As the College Board changed test content to make it better for digital distribution, the digital SAT still tests the same skills and knowledge learned in high school that are most important for college and job readiness.

Digital SAT Vs. Previous SAT: What Has Not Changed?

Despite the fact that the digital SAT comes up with a number of changes that is beneficial to both students and colleges, many of the SAT's most essential components stay unaltered.

- The digital SAT still tests students on the skills and knowledge they need to be ready for college and careers.
- The SAT is scored on the same 1600-point scale, so teachers and students are able to see how much they have improved over time.
- The digital SAT cannot be taken at home but rather in a school or testing center in a proctored setting.

Does a student Need a Device to Take the Digital SAT?

The digital SAT is taken on a laptop or tablet. Students can use their personal or test center-issued devices. If a student does not have a laptop, they may request one from the College Board, and it is provided to them on their SAT Day. If a student has any device or connectivity issues on the test day, the College Board has designed the test application so that it survives internet outages. Students are able to continue with the test without interruption if the internet goes down; their responses are stored, and they will not lose any testing time.

How Challenging Is It to Take the SAT Online?

The good news is that most students who took the digital SAT found the new version of the test easier to understand and use. When the College Board says that the digital version of the test is more "intuitive," it implies that it is "easier to navigate" and "easier to answer without becoming exhausted."

The digital SAT interface has a provision for a clock, which makes it easier for students to keep track of time.

In addition to highlighters, strikethrough, and bookmarks for review, there is an integrated calculator and a provision for a standard math formula sheet. At a single location, test-takers find all the resources necessary to answer questions.

What Tools Are Available for Students Taking the Digital SAT?

The test application includes many testing tools. Here are two examples:

- You can mark questions for review using the bookmark feature.
- Test-takers have the option to display or hide the digital clock placed at the top of their testing screen.
- During the math section, students may use the integrated graphing calculator (or one can bring their own basic calculator).
- A reference sheet with standard mathematical formulas for each question is provided.

When and How to Study for the Digital SAT?

Start preparing for the digital SAT as quickly as possible. Focus on making noticeable gains in every area: reading, writing, and math. Don't be hesitant to monitor your progress using the previous SAT tests. There is no need to take full-length tests, but you may always attempt to earn a perfect score on practice tests. Before starting to read, emphasize the need to develop comprehension skills and knowledge of previous global issues and concerns. This will gradually boost your vocabulary and reading speed, preparing you for future SAT reading success.

How to Study for the Digital SAT?

Reading, writing, grammar, and math skills are still the most important parts of the SAT for measuring college readiness.

This is how to begin preparing for each section of the digital SAT:

- **Reading and Writing (RW)**: Start with a variety of humanities- and science-related texts. You should read the New York Times, the Economist, the New Scientist, and the Washington Post. You should start reading the editorials. Learn about contemporary environmental and geopolitical concerns. Focus on being a knowledgeable reader in general. Instead of being a passive recipient of information, develop views. Application of grammar and punctuation rules is still tested on the SAT.

- **Math**: Start practicing questions on all the math topics on the previous SAT. In addition, pay close attention to the geometry and trigonometry questions on the official SAT material and Manhattan Review guides. Learn the formulae and their application to the previous SAT topics, as well as the rationale behind their operation.

How Would These Changes Impact College Admissions?

Since the SAT's basic content, level of difficulty, and scoring system did not change much, we expect schools to evaluate the digital version in the same way they evaluate the previous version. In addition, colleges make an attempt not to fault students for situations beyond their control, so taking one version of the SAT will not negatively affect your college applications relative to those who took the other version.

Even with these changes, colleges have a very different view of the SAT and ACT than they had a few years ago. Several schools are paying less attention to standardized test scores and more attention to other parts of an application, such as a student's grade point average (GPA), the courses they took in high school, and extracurricular activities. The COVID-19 epidemic made this worse by making it difficult or impossible for many students to take standardized tests for several months. Almost every school made testing optional in 2020 and 2021, and hundreds of colleges have decided to stay test-optional permanently. If a school is test-optional, you have the choice of submitting SAT or ACT scores with your application. If you do not, the other application aspects will carry more weight.

The College Board is aware of these trends, and many of the changes to the SAT were made to make it easier for students who might have been turned off by the length and difficulty of the test. Still, a College Board survey showed that 83% of students wanted to be able to send test scores with their college applications, and almost all schools still let them, so the SAT is here to stay.

Key Takeaways

The following are the six most significant takeaways:

- The SAT is administered digitally.
- The duration of the SAT is reduced (to about 2 hours instead of about 3 hours).
- Students have more time to answer each question.
- Calculator is permitted for the entire math section of the test.
- Reading passages is of shorter length. Each passage has a single question.
- Students get their digital SAT scores within days rather than weeks.

Digital SAT Math content

Content	**Topics**	**Number of questions**
Algebra	Fundamental concepts used in Algebra; arranging formulae, linear equations, inequalities, etc.	13–15
Advanced Math	Advanced concepts in Algebra, incl. absolute value, quadratic, exponential, polynomial, rational, radical, and other nonlinear equations	13–15
Problem Solving & Data Analysis	Understanding qualitative & quantitative data, analyzing relationship, incl. graphical; Ratio & Proportion, Percents, and units of measurements	5–7
Geometry & Trigonometry	Geometry (angles, triangles, area, and volume), Trigonometry (Trigonometric ratios)	5–7
	Total time (70 minutes)	**44**

Chapter 3

Practice questions

3.1 Algebra

3.1.1 Algebraic Expressions

1. What is the value of $(m-3)^3$ if $m = (\frac{5}{3} + \frac{5}{6} + \frac{1}{2})$?

(A) -1

(B) $-\dfrac{1}{8}$

(C) 0

(D) $\dfrac{1}{27}$

2. The value of Q is given by the relation $Q = 17m$. If Q and m are integers with Q greater than 150 and less than 160, then which of the following is the value of m?

(A) 9

(B) 10

(C) 12

(D) 14

3. The value of Q is given by the relation $Q = 17m$. If m is an integer greater than 15 and less than 23, then which of the following could be the value of Q?

(A) 255

(B) 290

(C) 306

(D) 391

4. If $N = \dfrac{S}{\frac{p}{q} - 1}$, then express q in terms of $N, S,$ and p.

(A) $q = \dfrac{Np}{N+S}$

(B) $q = \dfrac{Np}{N-S}$

(C) $q = \dfrac{N}{N+S}$

(D) $q = \dfrac{N}{Np+S}$

5. Calculate the value of $\frac{3}{\frac{3}{4}-\frac{5}{12}}$.

(A) $\frac{3}{4}$

(B) $\frac{9}{4}$

(C) 9

(D) 12

6. **SPR:**

If $p = \frac{1+\frac{x}{y}}{1+\frac{xy}{z^2}}$, $x = \frac{z}{2}$ and $y = 3z$, find the value of p.

7. **SPR:**

If $a = 216$, find the value of $a^{\frac{1}{3}} - 1$.

8. **SPR:**

What is the value of $\left(\frac{1}{x-9}\right)^n + \left(\frac{1}{x}\right)^{1-n}$ if $x = 25, n = \frac{1}{2}$?

9. **SPR:**

What is the value of $\frac{a^2+b^2}{(a+b)^2+(a-b)^2}$ if $a = 2.3$, and $b = 1.7$?

10. Find the value of the expression $\frac{x^2-12x+32}{x-4}$ if $x = 6.5$.

(A) -1.5

(B) 2.5

(C) 45.16

(D) 58.83

11. Find the value of the expression $\frac{x^2-12x+32}{x+4}$, when $x = -6.5$.

(A) -60.5

(B) 2.5

(C) 45.16

(D) 60.9

12. SPR:

Let # be an operation on x and y defined as $x\#y = \frac{x^2+y^2}{x+y}$. Find the value of 1#1#5 given that the operations are performed from left to right?

13. SPR:

If a mixed fraction is denoted by $Z\frac{Z}{3} = \frac{8}{3}$, then what is the value of $\frac{1}{2Z-3}$?

3.1.2 Numbers

14. Which of the following must be odd?

I. The sum of 5 consecutive integers

II. The sum of 14 consecutive integers

III. The product of 11 consecutive integers

(A) Only I

(B) Only II

(C) Only III

(D) Both I and II

15. Three consecutive numbers are selected from a set of integers from 1 to 30. If P is the product of the numbers drawn, which of the following must be true?

I. P is an integer multiple of 3.

II. P is an integer multiple of 4.

III. P is an integer multiple of 6.

(A) Only I

(B) Only II

(C) Both I and III

(D) Both II and III

16. If $a, b,$ & c are single digit integers and $N = 11,000 \times a + 10 \times b + c$, find the digit in the hundreds place of number N.

(A) 0

(B) 1

(C) a

(D) b

17. **SPR:**
How many two digit numbers exist such that the numbers are 54 greater than their reversed numbers?

18. **SPR:**

Let # be a digit from 0 to 9 such that $\frac{1\#}{3\#} + \frac{4}{7} = 1$.

(Note: Here 4# and 9# represent two digit numbers).

Find the value of the digit #.

19. Bob writes down all consecutive integers starting from 1 till 100 and adds them to get a number P. Alfred writes down all consecutive even numbers starting with 2 till 200 and adds them to get a number Q. Choose the correct relation between P and Q.

(A) $P = Q$

(B) $P = 2Q$

(C) $P = \frac{Q}{2}$

(D) $P = Q^2$

20. If $2a = 5b = 3c = 6d$ for four positive integers $a, b, c,$ & d, which of the following definitely does NOT represent an integer?

(A) $\frac{ad}{3}$

(B) $\frac{ac}{d}$

(C) $\frac{ac}{bd}$

(D) $\frac{ab}{cd}$

21. If $ab = cd$ for four positive integers $a, b, c,$ & d, which of the following must NOT be true?

(A) $\frac{ad}{c} = \frac{d^2}{b}$

(B) $\frac{a^2 + c^2}{c^2} = \frac{d^2 + b^2}{b^2}$

(C) $\frac{\frac{1}{b} + a}{a} = \frac{\frac{1}{c} + d}{d}$

(D) $\frac{ab^2}{c} = d$

22. A crate contains green and red apples in the ratio 7 : 11. When 10 green apples and 10 red apples are removed from the crate, the ratio becomes 9 : 17. How many red apples were originally in the bag?

(A) 10

(B) 12

(C) 21

(D) 28

23. **SPR:**
The symbol @ is defined such that $a@b = a^2 - b^2$. If $x@4 = \frac{2x+8}{5}$, find the value of x given that $x \neq -4$.

24. How many three-digit numbers yield remainder 1, when divided by 100 and by 60?

(A) One

(B) Two

(C) Three

(D) Four

25. If u and v are integers and $uv = -7$, what is the value of $(u + v)^2$?

(A) -36

(B) 4

(C) 36

(D) 49

3.1.3 Factors, Multiples, LCM, & HCF

26. How many factors of 48 are greater than $\sqrt{48}$?

(A) Four

(B) Five

(C) Six

(D) Eight

27. **SPR:**
A rectangular floor of dimensions 48 feet by 60 feet is to be tiled, using identical square tiles. What is the minimum number of square tiles required?

28. Amy decides to write down every fourth number starting from 1 till 50, with the first number being 1. While Mary decides to write down every third number starting from 1 till 50, with the first number being 1. How many numbers would be common to both of them?

(A) Three

(B) Four

(C) Five

(D) Six

29. After multiplication with 5, which of the following four numbers does not have the same number of distinct prime factors as the other three?

(A) 15

(B) 28

(C) 30

(D) 22

30. Which of the following numbers, given in options, has the highest exponent for any of its prime factors?

(A) 48

(B) 72

(C) 250

(D) 484

31. If $75^n \times 15 = 3^6 \times 5^m$, find the value of $(m + n)$ if m and n are positive integers.

(A) 5

(B) 8

(C) 11

(D) 16

32. If $1,500 = 6^m \times 50^n \times 5$, find the value of $(m - n)$ if m and n are positive integers.

(A) 0

(B) 1

(C) 2

(D) 3

3.1.4 Linear equations

33. If $x + y = 21$ and $2x = 5y$, find the value of $\sqrt{x - y} + (\frac{x}{y})^2$.

(A) 3.16

(B) 9.16

(C) 9.25

(D) 15.25

34. The cost of 75 apples and 45 oranges is \$12. At the same rate for each fruit, what would be the price of 50 apples and 30 oranges?

(A) \$4

(B) \$8

(C) \$18

(D) \$20

35. **SPR:**

A man bought some apples and some mangoes. The number of apples he bought was 4 less than twice the number of mangoes. Had he bought 6 mangoes more, the number of apples and mangoes would have been equal. How many mangoes did the man buy?

36. In the city Xuangho, fine for speeding on the highway is calculated in a strange manner. There is a flat fine of \$60 for anyone found driving above the speed limit of 100 miles/hour and an additional fine of \$3 for each mile/hour if found driving above the limit of 120 miles/hour. Lucy was fined \$90. At what speed (in miles/hour) was Lucy driving?

(A) 90

(B) 110

(C) 130

(D) 190

37. Seven thieves steal x gold coins from a treasury, where x being an even number. Upon trying to distribute the coins equally among themselves, they find that there are 6 coins left over. How many coins, when added to x, may result in a number of coins divisible among 14 thieves?

(A) 13

(B) 23

(C) 36

(D) 42

38. Presently, Amy is six years younger to Jane. Five years ago, Jane was thrice as old as Amy. What will be Jane's age ten years from now?

(A) 8

(B) 14

(C) 18

(D) 24

3.1.5 Absolute numbers

39. If $|5x| + 30 = 40$, what is x?

(A) -2

(B) 2

(C) ± 2

(D) $\sqrt{2}$

40. What is the value of $\left[\frac{-|-25| \times (-|2|)}{(-|-5|)^2}\right]$?

(A) -2

(B) 2

(C) $-\sqrt{2}$

(D) $\sqrt{2}$

41. If $\frac{2x}{3y} = 0$, which of the following MUST be true?

(A) $y \neq 0$

(B) $y = 0$

(C) $x \neq 0$

(D) $x < y$

42. **SPR:**

If $|3x - 4| = 11$, what is x?

43. **SPR:**

If $\frac{|13x - 4|}{5} + 3 = 10$, what is x?

3.1.6 Inequalities

44. If $|2x + 1| < 7$, what is the range of value of x?

(A) $4 > x > -3$

(B) $-4 > x > 3$

(C) $x < 3$

(D) $-4 < x < 3$

45. If $\frac{1}{x-1} > 1$, what is the range of value of x?

(A) $0 > x > -2$

(B) $0 < x < 1$

(C) $0 < x < 2$

(D) $1 < x < 2$

46. If $\frac{-2|x|}{3y} < -1$, which of the following MUST be true?

(A) $x > 0$ & $y > 0$

(B) $|x| < 3$

(C) $y > 0$

(D) $x < y$

47. If $-3 < -x < -1$, which of the following could be true?

I. $x^2 < 2x$

II. $x^2 = 2x$

III. $x^2 > 2x$

(A) I only

(B) II only

(C) III only

(D) All three

48. If $a^3bc^2 < 0$, which of the following must be correct?

(A) $a < 0$ & $c < 0$

(B) $b < 0$ & $c < 0$

(C) $a < 0$ & $b > 0$

(D) $a < 0$ & $b > 0$ or $a > 0$ & $b < 0$

49. SPR:

If $6 > x > 3$ and $-4 < -y < 7$, how many integer of values can $(3x + 2y)$ have?

3.1.7 Graphic equations & Inequalities

50.

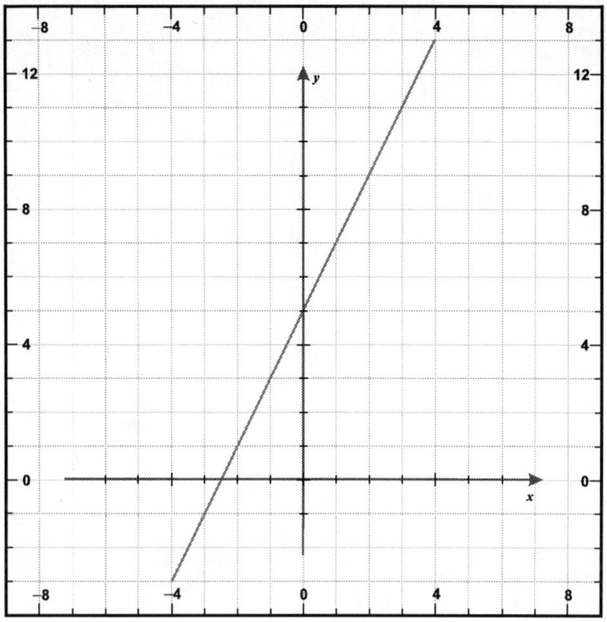

The graph of $f(x) = 2x + 5$ is shown above. Choose from the options below the correct graph of $g(x) = 2(x - 1) + 5$.

(A)

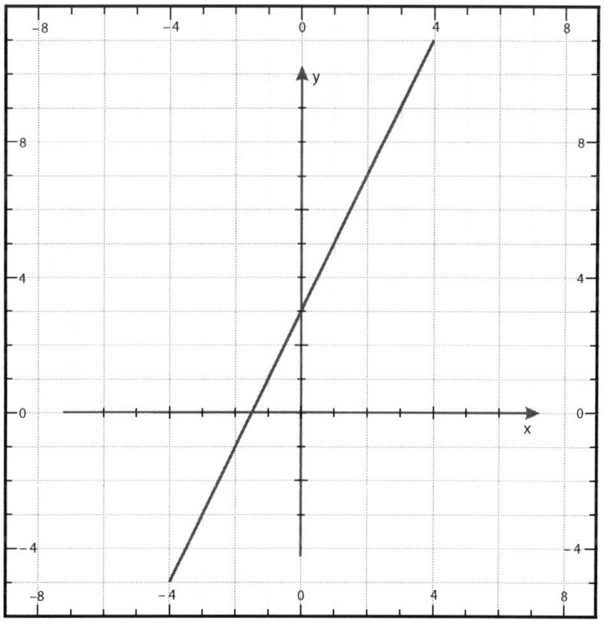

(B)

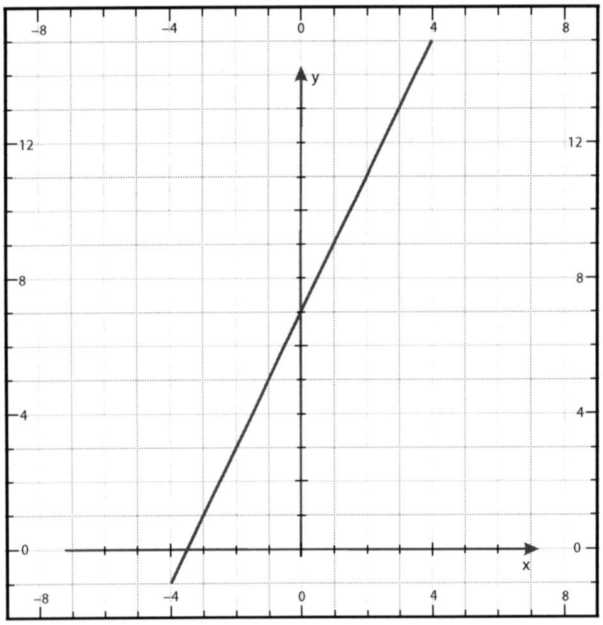

(C)

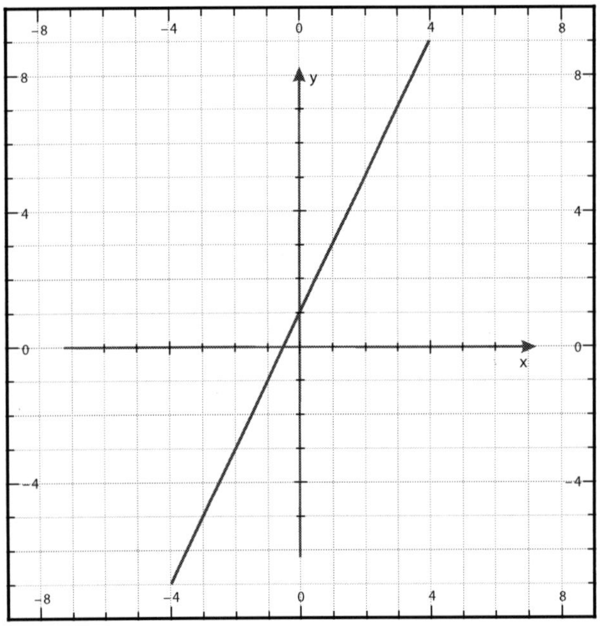

(D)

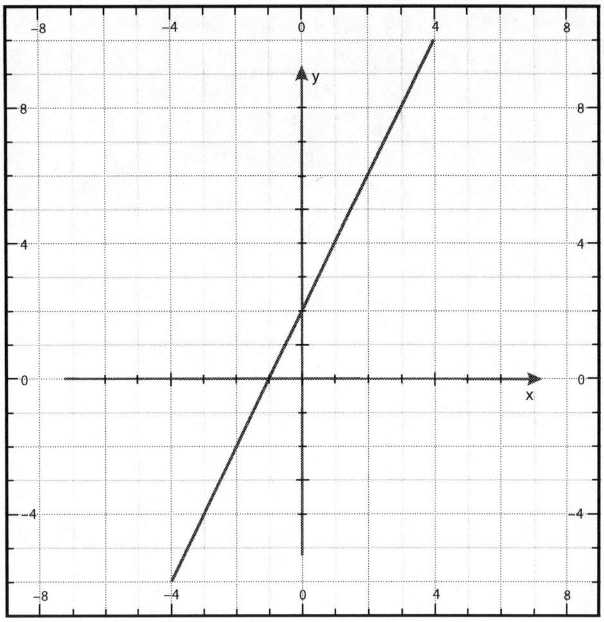

51.

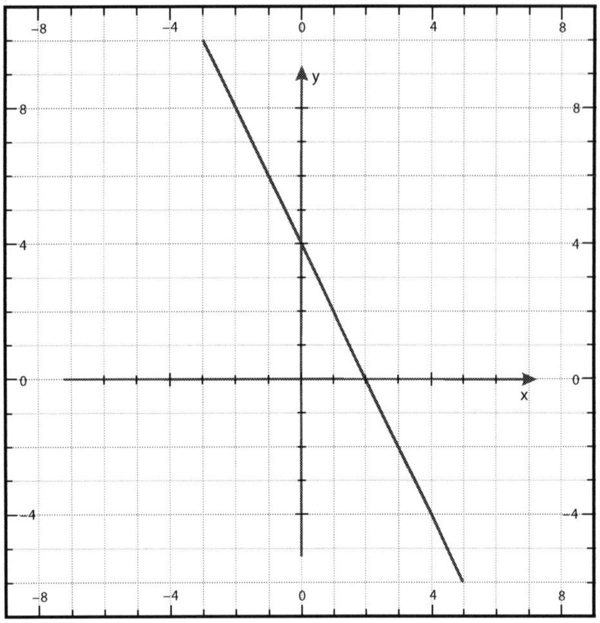

$y = f(x)$

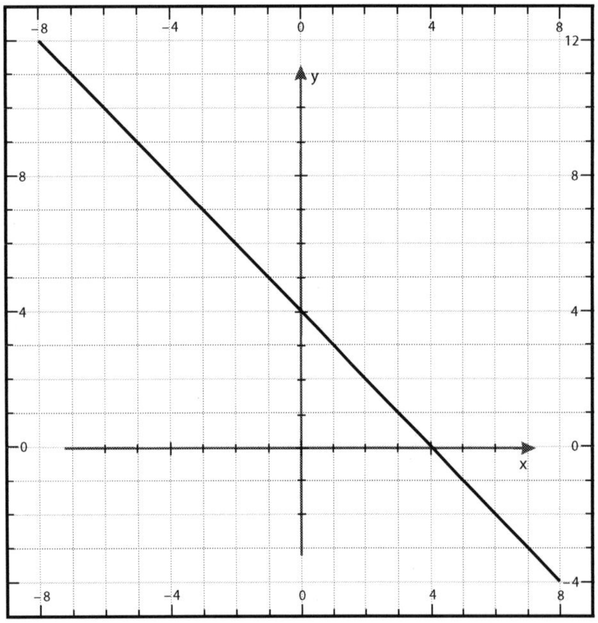

$y = g(x)$

The graphs of $f(x)$ and $g(x)$ are shown above. Choose from the options below the correct relation between $g(x)$ and $f(x)$.

(A) $g(x) = 2f(x)$

(B) $g(x) = \frac{f(x)}{2}$

(C) $g(x) = f(2x)$

(D) $g(x) = f\left(\frac{x}{2}\right)$

52.

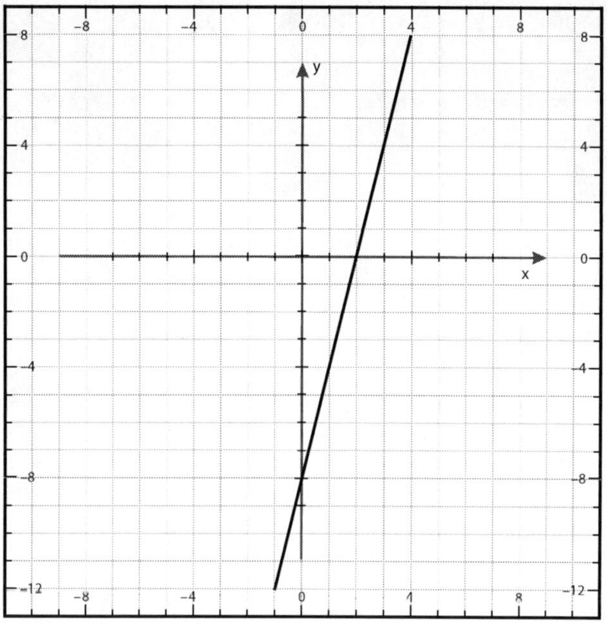

$y = f(x)$

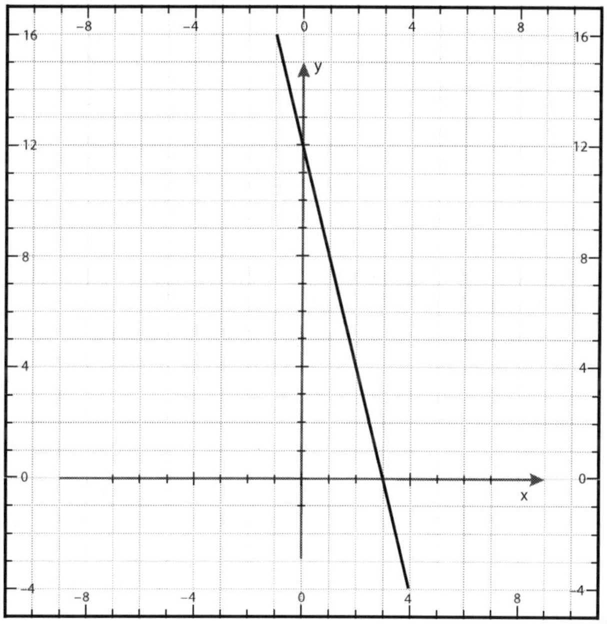

$y = g(x)$

The graphs of $f(x)$ and $g(x)$ are shown above. Choose from the options below the correct relation between $g(x)$ and $f(x)$.

(A) $g(x) = f(x) - 1$

(B) $g(x) = -f(x - 1)$

(C) $g(x) = -f(x) + 1$

(D) $g(x) = -f(2x) - 1$

53. Choose the graph from the options below that correctly depicts the region denote by $f(x) \geq 4|x - 2| - 3$ (the region has been shaded in the options).

(A)

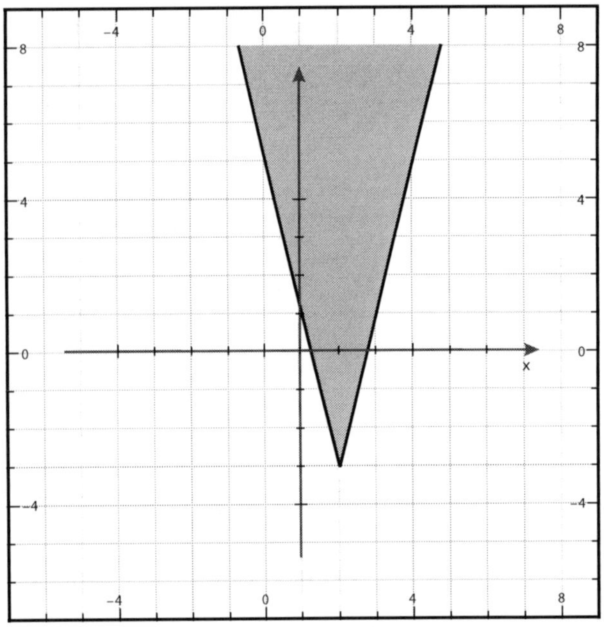

(B)

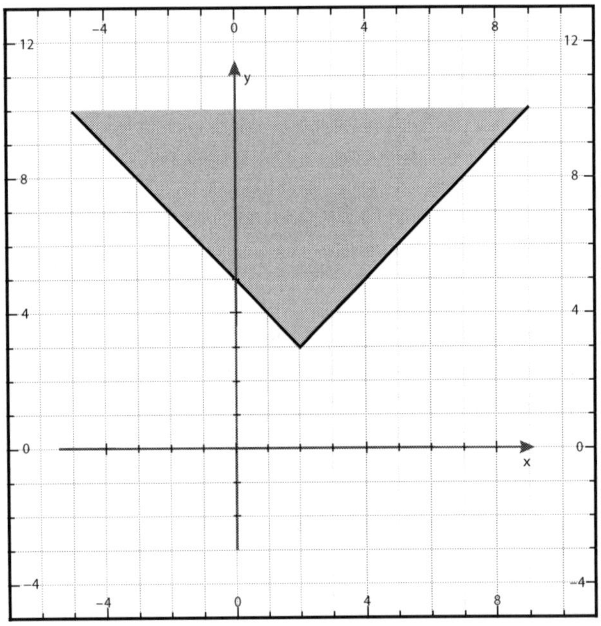

(C)

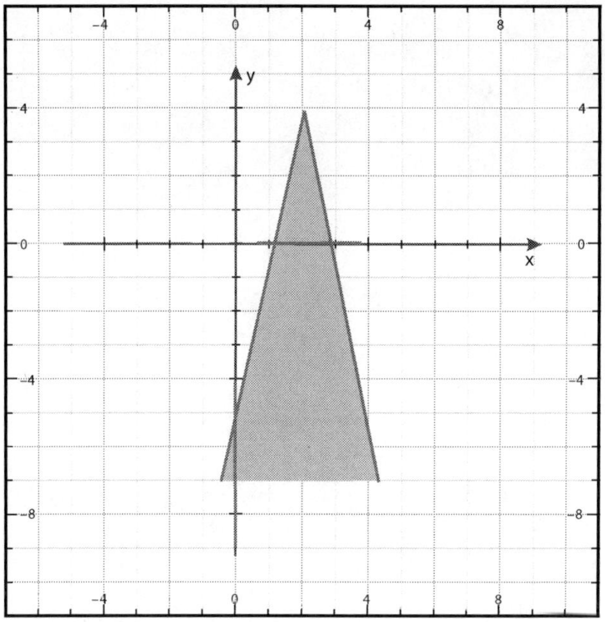

(D)

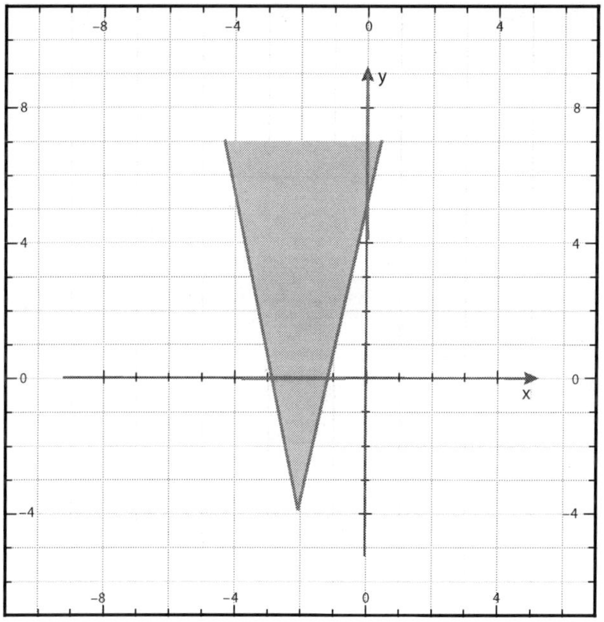

3.1.8 Functions

54. If $f(x) = 3x^3 - 2x^2 + 2x + 2$, what is the value of $f(-2)$?

(A) -34

(B) -14

(C) 14

(D) 25

55. If $g(t) = 2t^3 - 3t^2 + \frac{3}{t^2} + \frac{2}{t^3}$, what is the value of $g(-\frac{1}{t})$.

(A) $-g(t)$

(B) $g(t)$

(C) $-2g(t)$

(D) $2g(t)$

56.

$f(x)$	3	2	10	23
x	0	1	−1	−2

Based on the table above, which of the following could represent $f(x)$?

(A) $f(x) = -4x^2 + 3x + 3$

(B) $f(x) = -5x^2 + 4x + 3$

(C) $f(x) = 3x^2 - 4x + 3$

(D) $f(x) = -4x^2 - 3x + 3$

57. SPR:

Given $f(x) = 3x - 4$, for what value of x, does $5f(x) - 3 = f(3x - 4)$?

58. SPR:

If for all positive integers $m, [m] = -|\sqrt{m}|$, when m is odd & $[m] = -\sqrt{m}$, when m is even; and for all negative integers $m, [m] = m^2$, when m is odd & $[m] = -|m^2|$, when m is even, what is the value of $\frac{[49]^4 \times [-4]}{[-49] \times [4]}$?

59. If $[X]$ is the smallest integer greater than or equal to X, what is the value of $[-2.2] + [3] + [5.2]$? Here a dot (.) is a decimal.

(A) 5

(B) 6

(C) 7

(D) 8

60. An operation @ is defined by the equation $a@b = \frac{(a+b)}{(a-b)}$, for all numbers a and b such that $a \neq b$. If $a@c = 2$, then what is the value of c in terms of a?

(A) $-a/3$

(B) $-3a$

(C) $a/3$

(D) $3a$

3.2 Problem Solving & Data Analysis

3.2.1 Time & Work

61. Amy and Bob decide to paint one wall of a building. Working alone, Amy takes 12 hours to paint the entire wall, while Bob takes 18 hours for the same. Amy painted the wall for 4 hours and then Bob took over and completed the wall. How long did it take for them to paint the entire wall?

(A) 12

(B) 16

(C) 27

(D) 31

62. **SPR:**
John can do a piece of work in 20 days. He worked initially for 12 days after which he asked his friend Matt to complete the remaining work for him. The work required to be done by Matt is what percentage less than the work already done by John?

63. Joe takes 6 hours to water plants in his garden. When Joe works with his brother Jack, they take 4 hours to water the plants. How many hours would Jack take, working alone, to water the plants?

(A) 1.2

(B) 2.0

(C) 10

(D) 12

3.2.2 Mixtures

64. A 50 ml. mixture of wine and water has only 15 percent wine. If a person requires 27 ml. of wine, what volume of the mixture must he have access to?

(A) 81

(B) 108

(C) 150

(D) 180

65. On average, 50 grams of a certain metal makes only 15 percent of the total yield from an ore. If a technician observes 27 grams of this metal in a sample of an ore, what percent of the total yield does it make?

(A) 4.5%

(B) 5.4%

(C) 8.1%

(D) 27.78%

66. The ratio of quantity of milk to water in a mixture was 1 : 3. If 60 liters of milk was added to the mixture, the ratio of quantity of milk to water in the mixture became 3 : 5. What was the total quantity of the initial mixture?

(A) 120

(B) 135

(C) 225

(D) 300

67. An energy drink concentrate widely available in the market contains 10% carbohydrates, 40% protein, 30% glucose and 20% minerals. John buys this drink and mixes it with water so that water forms 60% of the mixture. What percentage of the new mixture thus formed is constituted of minerals?

(A) 4%

(B) 8%

(C) 12%

(D) 16%

68. Alfred, a tour operator, wanted to increase his fleet of cars. For that, from his customers, he wanted to know the preference of car types. A survey revealed that 40% of the customers preferred sedan, while the remaining preferred SUV. Of those who preferred sedan, 45% preferred Volkswagen while 55% preferred Honda. What percentage of the cars should Alfred have Honda sedans so that it is commensurate with the findings of the survey?

(A) 18%

(B) 22%

(C) 40%

(D) 55%

3.2.3 Word Problems

69. **SPR:**

An airline has a policy of charging for baggage carried by its customers. It charges $\$x$ for luggage up to 5 lbs. For every lb of luggage above 5 lbs, it charges $\$y$ extra. A customer paid $\$a$ for his luggage. If it is known that $a > x$, and the weight of luggage carried by the customer in an integer, then find the expression that gives the weight of luggage carried by the customer.

(A) $5 + a - x$

(B) $5 + \dfrac{a - x}{y}$

(C) $5 + \dfrac{a}{y}$

(D) $5 + \dfrac{x}{y}$

70. **SPR:**

A man's salary increases from $\$x$ per month to \$400 per month as a result of a 25% hike. Find the value of $(x - 100)$.

71. **SPR:**

A manufacturing firm manufactures screws, rivets and nuts. The average times taken to manufacture a screw, a rivet and a nut are 10 seconds, 20 seconds and 80 seconds, respectively. If the firm manufactures screws from 10 AM to 11 AM, rivets from 11 AM to 12 noon, and nuts from 12 noon to 2 PM, what percentage of the total items manufactured are nuts?

72. Company X manufactured three new products P, Q, and R. To promote them among its three dealers, it wants to distribute 22 items of P, 37 items of Q, and 13 items of R such that each dealer gets an equal number of items. If each dealer gets at least 2 items of each product, what could be the maximum number of items of R stocked with any dealer?

(A) 2

(B) 9

(C) 11

(D) 20

73. The operating cost of a bus consists of a fixed maintenance cost calculated on a per day basis and fuel cost calculated at the rate of $\$n$ per mile travelled. The bus operates 6 days a week, making 3 trips per day covering a distance of d miles per trip. The total operating cost of the bus for an entire week is $\$c$. How much is the daily maintenance cost ($) of the bus?

(A) $(c - 3dn)$

(B) $\frac{(c - 3dn)}{7}$

(C) $\frac{(c - 18dn)}{6}$

(D) $\frac{(c + 6dn)}{2}$

74. Of 4,500 items on an assembly line, 30% turn out to be defective and are sent for rework. Of these, 80% of them are rectified and sent for processing and the rest are scrapped. How many more items are rectified after rework than those which are scrapped?

(A) 270

(B) 810

(C) 1,080

(D) 3,420

75. A freezer has a powerful ice forming system; it can form 2 inches of ice in every 40 minutes. The freezer was kept operational for 4 hours and after that it was switched off for 2 hours. Once switched off, the ice melts at the rate of 0.5 inches per hour. The freezer was again switched on for 2 hours. What is the total thickness of ice now?

(A) 8

(B) 11

(C) 17

(D) 18

76. SPR:

In a charity organized by a school, books were given away to children. Of them, $\frac{3}{11}^{\text{th}}$ were given away to children less than 8 year old. Of the remaining, $\frac{3}{4}^{\text{th}}$ were given to children aged between 8 and 12 years. The rest were given away to children older than 12 years. What fraction of the total books was given away to children older than 12 years?

77. SPR:

The number of student enrollments for a mathematics course in a college reduced from 235 in the year 2014 to 200 in the year 2015. What is the percentage reduction in the enrollment of students from 2014 to 2015?

78. **SPR:**

To manufacture an item, Process A requires assembling of 540 components. A machine can manufacture 60 such components in 25 minutes. To manufacture the same item, Process B requires assembling of 300 such components. A machine can manufacture 50 such components in 15 minutes. How much manufacturing time in percent can be saved if Process A is replaced by Process B (assume that assembling the components happens instantly)?

79. In Singapore, in the year 2011, the number of people who voted increased by 2.4 million and the number of people eligible to vote increased by 10.8 million. For 2011, find the percentage of people who voted of people who were eligible to vote if in 2010, the number of people eligible to vote and the number of people who had voted were 192 million and 54 million, respectively.

(A) 22.22%

(B) 27.81%

(C) 28.12%

(D) 29.38%

80. To gain entry for the Bachelor's programs in different colleges, students need to appear for the AST entrance test. Of all students who register for the test, only 80% appear for the test. Of them, only 75% clear the test. By what percentage is the number of students after registering for the test do not appear for the test less than the number of students who clear the test?

(A) 33.33%

(B) 40%

(C) 60%

(D) 66.66%

81. To gain entry for the Bachelor's programs in different colleges, students need to appear for the AST entrance test. A total of 50,000 students register for the test in any given year. Of all students who register for the test, only 80% appear for the test. Of these students, only 75% clear the test and gain entry into a college of his/her liking. What is the difference between the number of students who do not appear for the test after registering and the number of students who clear the test?

(A) 10,000

(B) 20,000

(C) 30,000

(D) 40,000

82. A man purchased goods at the rate of 12 for \$5. If he sold them at the rate of 10 for \$6, what is his percent profit?

(A) 16.66%

(B) 20%

(C) 30.55%

(D) 44%

83. Running advertisements on prime-time television costs \$12,000 per 10 seconds. Running the same advertisement early morning reduces the cost by 30%. What is the total daily expenditure (in thousands of dollars) if a 30-second advertisement is to be run once during prime-time and twice early morning?

(A) 28.80

(B) 51.00

(C) 86.40

(D) 97.20

84. A sprinter covers d feet up a steep slope at a speed of s feet/second. While running down the slope, she followed the exact same route and found that she was able to run at a much faster pace of $(2s + 1)$ feet/s. What was the average speed (in feet/second) of the sprinter for the entire round trip?

(A) $\dfrac{\sqrt{s(2s+1)}}{2}$

(B) $\dfrac{(3s+1)d}{2s(2s+1)}$

(C) $\dfrac{4s+1}{2}$

(D) $\dfrac{2s(2s+1)}{3s+1}$

85. A sprinter covers some distance up a slope at a speed of 3.6 feet/second. While running down the slope, she followed the exact same route and found that she was able to run much faster at 6 feet/second. She took a total of 6 minutes for her round trip. What was the one way distance covered by the sprinter?

(A) 750 feet

(B) 810 feet

(C) 864 feet

(D) 1,620 feet

86. A car traveling from San Francisco to Los Angeles travelled the first d miles at an average speed of 60 miles/hr and the remaining 60 miles it covered in 6 hours. What is the average speed (in miles/hr) of the car for the entire journey?

(A) $\frac{60(d+60)}{d+360}$

(B) $\frac{60(d+60)}{d+6}$

(C) $\frac{d}{60}+10$

(D) 35

87. **SPR:**

In a certain class, 40 students play football and 30 students play rugby. If 10 students play both games and 20% of the students in the class do not play any of the two games, how many students are there in the class?

88. At a gas station, cars arrive for fueling every p minutes during the peak hours from 9:00 am to 1:00 pm and every q minutes during the non-peak hours from 3:00 pm to 6:00pm. If cars do not arrive at the gas station at any time other than those mentioned above, what is the average number of cars that arrive at the station every minute?

(A) $\frac{4q+3p}{12pq}$

(B) $\frac{4p+3q}{7pq}$

(C) $\frac{4q+3p}{7pq}$

(D) $\frac{7pq}{4q+3p}$

89. **SPR:**

A fruit seller stocks apples, oranges, mangoes and guavas. The number of apples with him is one-third the number of mangoes and the number of guavas is one-fourth the number of oranges. If the number of mangoes is half the number of guavas, what fraction of the total fruits are mangoes?

90. **SPR:**

A shopkeeper sells CDs, Pen-drives and DVDs. The number of CDs with him is one-third the number of Pen-drives and the number of DVDs with him is one-fourth the number of Pen-drives. What fraction of the total items are DVDs?

91. **SPR:**
A container used in large container ships has dimensions of $4 \times 6 \times 20$ cubic meters. It is usually filled with smaller boxes which have dimensions of $4 \times 3 \times 2$ cubic meters. What is the maximum number of smaller boxes that can fit in a container?

92. A bicycle rider calibrated his distance measuring device at a shop at the entrance to the bridge. On the bridge, due to vibration of the bridge, the device malfunctioned and underestimated the distance by 10 centimeters for every 15 meters traveled. What was the distance in meters shown by the device at the end of the bridge, if the length of the bridge is known to be 1,800 meters?

(A) 1,704

(B) 1,710

(C) 1,779

(D) 1,788

93. **SPR:**
A Wall Street firm hired 140 students with an MBA degree and 40 students with an MA degree. The cost of hiring a student with an MA degree is half of that with an MBA degree. What percent of the total cost of hiring is the cost of hiring the MA students?

94. During t_1 hours in the morning, customers arrive at a bank at the rate of r_1 per hour, and during t_2 hours in the afternoon, customers arrive at a bank at the rate of r_2 per hour. What is the average rate of arrival in customers per hour to the bank during the whole day?

(A) $\dfrac{r_1 + r_2}{\frac{r_1}{t_1} + \frac{r_2}{t_2}}$

(B) $\dfrac{r_1 + r_2}{t_1 + t_2}$

(C) $\dfrac{r_1 t_1 + r_2 t_2}{t_1 + t_2}$

(D) $\dfrac{t_1 + t_2}{r_1 t_1 + r_2 t_2}$

3.2.4 Profit & Loss

95. A dealer bought a bicycle for \$120 and later sold it to a customer for \$90. What was the percent loss in the transaction?

(A) 10.0%

(B) 25.0%

(C) 30.0%

(D) 33.3%

96. To attract more customers, a distributor provides a 20% discount on the listed price of a product yet maintains a 10% profit. If the price at which the distributor procured the product is \$120, what should be the list price of the product?

(A) \$100

(B) \$125

(C) \$130

(D) \$165

97. By selling 16 toffees for \$4, a man makes neither profit nor loss. How many toffees should he sell for \$10 to make 25% profit?

(A) 32

(B) 36

(C) 40

(D) 50

98. A shopkeeper sells balloons at 10 cents apiece or at \$2 per box if purchased in boxes containing 25 balloons each. How much does a customer save if he purchases 500 balloons by purchasing boxes of balloons than by purchasing in single pieces?

(A) \$5

(B) \$10

(C) \$40

(D) \$50

99. A shopkeeper sells balloons either in single pieces priced at 10 cents apiece or in boxes containing 25 balloons at \$2 per box. What is the minimum amount that a customer has to spend if he wants to purchase exactly 310 balloons?

(A) \$12.8

(B) \$25.6

(C) \$25.0

(D) \$26.0

100. **SPR:**
A pen costs thrice as much as a pencil. A boy purchased 10 pens and 10 pencils. His total cost is \$20. What is the cost (in cents) of 5 pencils?

101. **SPR:**
A pencil costs 50 cents and a pen costs thrice as much as a pencil. With \$20, a boy wants to purchase a few pens and a few pencils. What is the maximum number of items he can purchase if he has to purchase at least 3 pens?

102. **SPR:**
A shopkeeper sets the selling price of a DVD such that the selling price of 7 DVDs equals the cost price of 9 DVDs. Given that a customer buys 9 DVDs, what percentage profit does the shopkeeper make?

103. **SPR:**
A shopkeeper sets the selling price of a DVD such that the selling price of 7 DVDs equals the cost price of 9 DVDs. Given that a customer buys 10 DVDs, and saves \$30, what is the cost price (in \$) of each DVD?

104. **SPR:**
A restauranteur promotes his restaurant by offering a scheme: "Dine 5, and Pay for 4". If a group of 5 dines in the restaurant, what percent of discount will they get?

105. **SPR:**
A shopkeeper promotes his shop by offering a scheme: "Buy 5, and get 3 free". If a man wants 8 shirts, what percent of discount will he get?

106. When Brian bought a boat, he found out that the municipality charged him 7 percent tax for marina services. If Brian paid \$6, 885.45 for the boat, what was the price of the boat before marina services tax?

(A) $6,012.34

(B) $6,535.00

(C) $6,435.00

(D) $7,367.43

107. Initially, the price of an item had been expressed in terms of whole dollars. A store raised the price of an item by exactly 10 percent. Which of the following CANNOT be the resulting price of the item?

(A) $5.50

(B) $6.60

(C) $11.70

(D) $75.90

108. **SPR:**

Michael runs a book shop. For every customer who buys four books, he sells the first book at the listed price and gives 10% discount on each of the other three books. Assuming each book is priced at $60, what is the net amount (in dollars) paid by a customer who buys four books?

3.2.5 Interest & Compounding

109. SPR:
In a laboratory experiment on bacteria, it was observed that the count of bacteria doubles in every 60 minutes. In order to control their growth, scientists introduced a chemical agent, which resulted in the bacteria multiplying to thrice their count in two hours. By what percentage was the growth of the bacteria reduced?

110. An investor puts in \$1,000 for a year with a financial institute. Which of the following terms of payment of interest would be most lucrative to the investor?

(A) Interest calculated at the end of the year at the rate of 10% per annum

(B) Interest calculated and compounded at the end of every 6 months at the rate of 9% per annum

(C) Interest calculated and compounded at the end of every 3 months at the rate of 8% per annum

(D) All the above terms of payment of interest would yield the same interests

111. An investment pays at the rate of $6\frac{1}{4}\%$ on the annual basis. If the same investment, instead of $6\frac{1}{4}\%$ on a annual basis, pays at an annual interest rate of 6% on a semi-annual basis without compounding, what is the gain or the loss in payments as per the new terms at the end of the year to the investor who invests 20,000 dollars?

(A) Loss of \$32

(B) Loss of \$50

(C) Gain of \$18

(D) Gain of \$50

112. During a festive season, an e-commerce website offers two schemes on buying iphones. The first offer demands an outright payment of 20% of the phone price and 5 equal monthly installments, with each being 20% of the phone price, while the second offer demands an outright payment of 10% of the phone price and 12 equal monthly installments, with each being 10% of the phone price. By what percent is the first offer cheaper than the second?

(A) 7.69%

(B) 8.33%

(C) 11.11%

(D) 20%

113. If the bacteria count of a culture increases by 10% per minute, during what interval will the bacteria count be more than 1.5 times the initial bacteria count?

(A) 2-3 minutes

(B) 3-4 minutes

(C) 4-5 minutes

(D) 5-6 minutes

3.2.6 Ratio & Proportion

114. To make a dye, one needs to have red and yellow paints in the ratio 3 : 5. How many liters of the dye can be made if there are 13.5 liters of red paint and 18 liters of yellow paint?

(A) 10.8

(B) 22.5

(C) 28.8

(D) 36.0

115. To make a perfume, a manufacturer needs to mix oil, fragrance and water in the ratio 1 : 8 : 3 by volume. What maximum quantity of perfume can the manufacturer make, using 60 ml. fragrance? Given that oil and water are abundant.

(A) 15 ml.

(B) 40 ml.

(C) 75 ml.

(D) 90 ml.

116. Two cities A and B are connected by three routes R, S, and P, whose lengths are in the ratio of 3 : 4 : 3. Three people travel from A to B, one in a SUV, second in a sedan, and third in a bike along the routes R, S, & P respectively. The amounts of fuel spent by the SUV, sedan and bike are in the ratio 6 : 2 : 3. What is the ratio of their fuel efficiencies i.e. fuel consumed per unit distance traveled by the SUV, sedan and bike?

(A) 4 : 1 : 2

(B) 1 : 4 : 2

(C) 18 : 8 : 9

(D) 3 : 2 : 2

117. A 40 liters mixture of milk and water had 80% concentration of milk. A part of the mixture was replaced with an equal quantity of water to achieve 50% concentration of milk. What volume of the mixture (in liters) was replaced to achieve the desired concentration?

(A) 12

(B) 15

(C) 20

(D) 32

118. A 40 liters mixture of milk and water had 80% concentration of milk. Some water was added to the mixture in order to achieve 50% concentration of milk. What is the volume of water (in liters) that was added to achieve the desired concentration?

(A) 12

(B) 24

(C) 32

(D) 64

119. In a laboratory experiment, a concentrated mixture containing 80% of a chemical is processed in a special container. 40% of this mixture is separated and stored for future use. The container is then filled to its full capacity by water. The chemical forms what percentage of the final mixture?

(A) 8%

(B) 12%

(C) 32%

(D) 48%

3.2.7 Permutation and Combination

120. 3-digit numbers are formed using the digits 2, 5, 8, and 9. Repetition of digits is allowed. How many of them are even?

(A) 8

(B) 16

(C) 32

(D) 64

121. How many different license numbers can be allotted using 3 letters and 3 digits? The letters and the digits may appear in any order. Repetition of the letters and the digits is allowed.

(A) $C_3^{26} \times C_3^{10}$

(B) $2 \times C_3^{26} \times C_3^{10}$

(C) $2 \times 26^3 \times 10^3$

(D) 36^3

122. How many 5-digit numbers can be formed containing exactly two 3s, appearing together? Other digits can be repeated.

(A) $3 \times C_3^9$

(B) $4 \times C_3^9$

(C) $9^4 + 32 \times 9^3$

(D) $9^3 + 24 \times 9^2$

123. In a polygon of 7 sides, how many different triangles can be drawn using the vertices of the polygon as the vertices of the triangles?

(A) 32

(B) 34

(C) 35

(D) 56

124. How many 3-member teams can be formed out of 6 individuals provided that one specific individual must be in the team?

(A) 5

(B) 10

(C) 15

(D) 20

125. In how many ways can 2-members: either only boys or only girls be chosen from a class comprising 8 boys and 10 girls?

(A) 28

(B) 73

(C) 153

(D) 1,260

126. How many distinct words (with or without meaning) can be formed using all the letters of the word: BANANA?

(A) 9

(B) 60

(C) 120

(D) 720

127. **SPR:** In how many ways can 2 girls or 2 boys be chosen from a class comprising 10 boys and 12 girls?

128. **SPR:** How many distinct words (with or without meaning) can be formed using all the letters of the word: “BANANA” such that the two “Ns" come together?

129. **SPR:** How many spy-codes are possible taking all symbols, together having the following shapes, if together □ must NOT be at the beginning and ◊ must NOT be at the end?

□ ⊵ ⊗ ⊘ ◊

3.2.8 Probability

130. A box contains 30 balls; of which 8 are black, 10 are green, and 12 are yellow. If two balls are drawn from the box at random, what is the probability that both balls will NOT be green?

(A) 3/9

(B) 24/29

(C) 8/9

(D) 26/29

131. A fair coin is tossed four times. What is the probability of getting at least one 'Tail'?

(A) 1/16

(B) 1/4

(C) 3/4

(D) 15/16

132. Jack, Parker, and Herman are given an assignment to collectively solve a problem. The probabilities of Jack, Parker, and Herman to solve the problem individually are 4/5, 2/3, and 3/4, respectively. What is the probability that the problem will be solved?

(A) 1/60

(B) 13/60

(C) 59/60

(D) 119/120

133. Three actors are to be randomly chosen out of five: Jack, Steve, Elad, Suzy, and Ali. What is the probability that if Jack is chosen in the three then Steve would be left out, and if Steve is is chosen in the three then Jack would be left out?

(A) 1/5

(B) 7/10

(C) 4/5

(D) 5/6

134. A box contains p number of silver coins and q number of gold coins. One coin is randomly drawn from the box but not replaced; thereafter a second coin is randomly drawn. What is the probability that both the coins drawn are silver?

(A) $\frac{pq}{(p+q)^2}$

(B) $\frac{p(p-1)}{(p+q)(p+q-1)}$

(C) $\frac{(p-1)(q-1)}{(p+q)^2}$

(D) $\frac{pq}{(p+q)(p+q-1)}$

135. A super mall issues coupons to its customers at the entrance of the mall. The coupons are numbered from 100 to 400, inclusive. Customers who have coupons with the numbers divisible either by '3' or by '5' win gifts. What is the probability that a random customer wins a gift?

(A) 141/300

(B) 141/301

(C) 161/301

(D) 161/300

136. A box contains 5 black and a few green balls. If two balls are drawn from the box at random, and the probability that both the balls are green is 1/6, how many green balls are in the box?

(A) 3

(B) 4

(C) 5

(D) 6

137. **SPR:** A pair of die is rolled. What is the probability of getting the sum of numbers greater than 10 on their upper faces?

138. **SPR:**

A coin is tossed three times. What is the probability of getting 'Tails' on the first two tosses and a 'Head' on the third toss?

139. SPR:

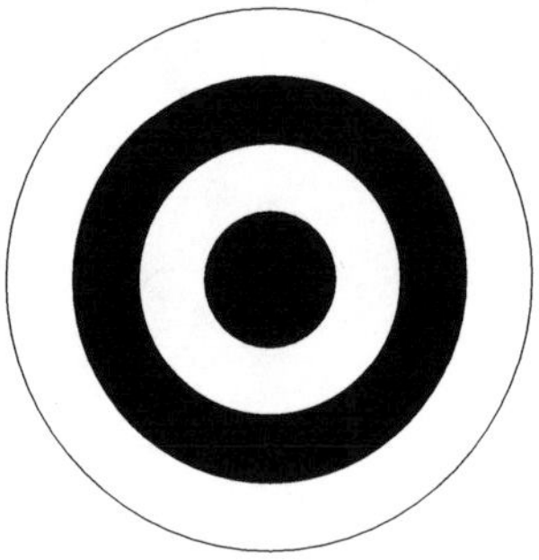

Above figure shows a dartboard, comprising four concentric circles. Radius of the the outermost circle is 4 inches, the radius of the innermost black circle is 1 inch, and the width of the black ring is 1 inch. If Brian throws a dart, and it hits the dartboard, what is the probability that it hits any of the black parts of the dartboard?

3.2.9 Statistics

Questions on Average (Arithmetic Mean)

140. Average (arithmetic mean) of x and y is 80, and the average of x and z is 200; what is the value of $\frac{(z-y)}{8}$?

(A) −30

(B) 0

(C) 30

(D) 80

141. Three numbers have their average (arithmetic mean) equals to 6. The middle number is 7. If the average of two larger numbers is 8, what is the average of the two smaller numbers?

(A) 2

(B) 4

(C) 4.5

(D) 6

142. The average (arithmetic mean) of six numbers is 7.5. If one of them is multiplied by 4, the average of the numbers decreases by 0.5, which of the following numbers is multiplied by 4?

(A) −3

(B) −1

(C) 0

(D) 2

143. A student's average (arithmetic mean) score on 5 tests is 38. What must be the student's score on the 6th test for the average score on all the 6 tests to be 41?

(A) 40

(B) 44

(C) 50

(D) 56

144. For the first x days, the average (arithmetic mean) rainfall was 32 millimeter per day. If today's rainfall of 56 millimeter increased the average rainfall to 36 millimeter per day, what is the value of x?

(A) 2

(B) 3

(C) 5

(D) 6

145. If all the elements of a set are multiplied by a constant, the average (arithmetic mean) remains unchanged, which of the following statements must true?

I. Average (arithmetic mean) = 0

II. Sum of the elements of the set equals 0.

III. The set contains at least one positive and one negative number

(A) I only

(B) II only

(C) I and II only

(D) I, II, and III

Questions on Median

146. If Set X: $\{-2, 23, 4, -8, 6\}$, and Set Y: $\{0, -2, -8, 8, 12, 13, y\}$ have equal median values, what is the value of y?

(A) 0

(B) 2

(C) 3

(D) 4

147.

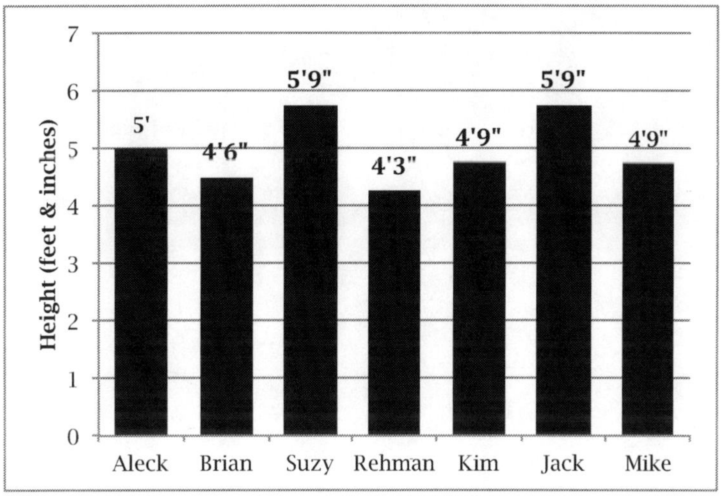

According to the given chart, which of the following student could represent the median height of the seven students?

(A) Aleck

(B) Suzy

(C) Kim

(D) Rehman

148. SPR:

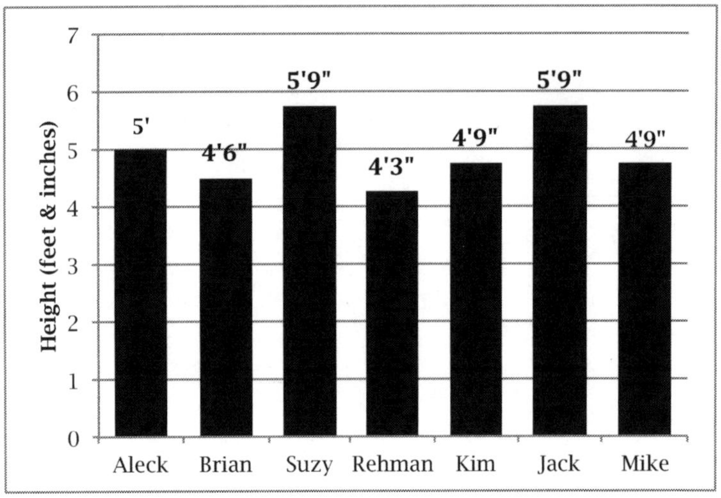

As per the data in the given chart, what is the average (Mean) height of the seven students in the nearest integer in centimeters?

149. SPR:

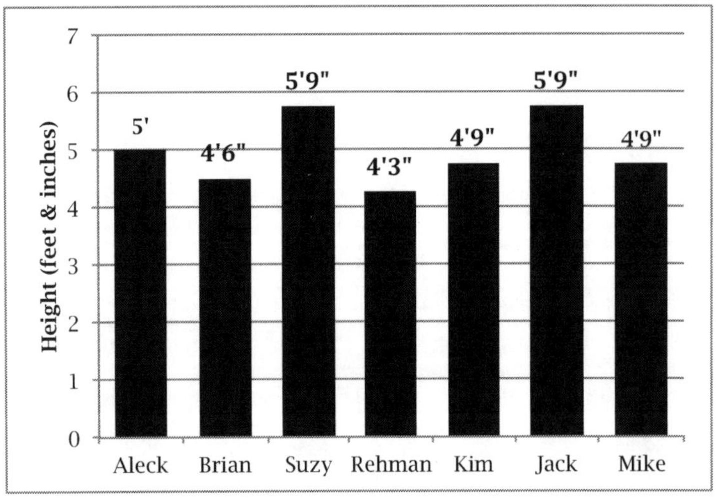

As per the data in the given chart, what is the range of heights of the seven students in inches?

150. If the median of X consecutive integers is 50.5, which is the smallest integer among the following?

(A) $51 - X/2$

(B) $50.5 - X/2$

(C) $50 - X/2$

(D) $49.5 - X/2$

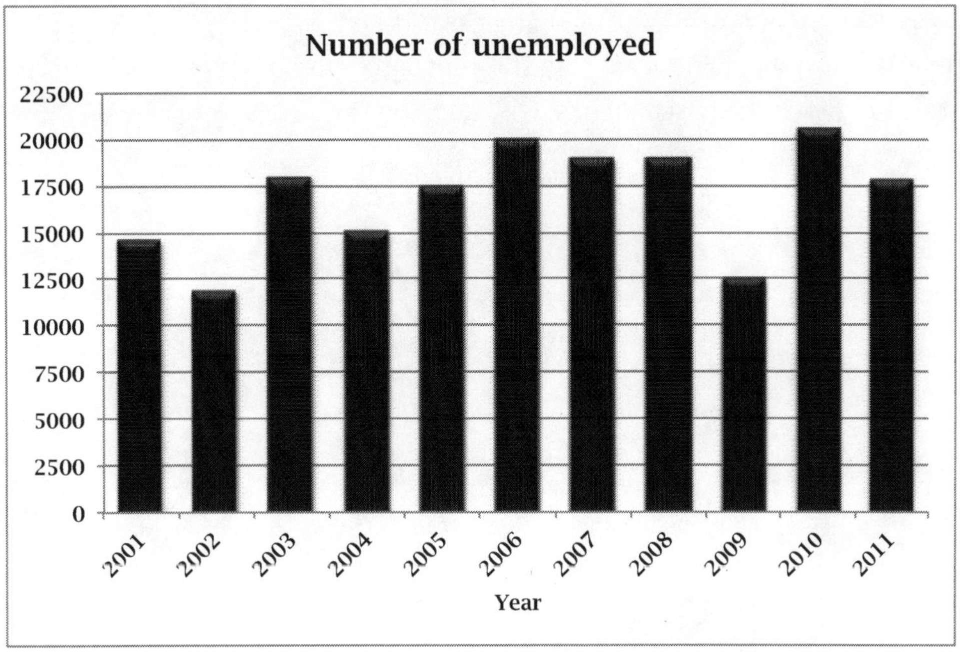

151. According to the chart given above, which of the following is closest to the median number of unemployed for the years 2001 to 2011, inclusive?

(A) 11,500

(B) 12,500

(C) 17,000

(D) 18,000

Questions on Average and Median

152. Number of defects in eight of the nine batches of batteries are: 2, 4, 6, 5, 3, 3, 7, 8, respectively. If the median defect per batch for nine batches equals the average (arithmetic mean) defect per batch for nine batches, which of the following would be the number of defects in the ninth batch?

(A) 2

(B) 3

(C) 4

(D) 7

153.

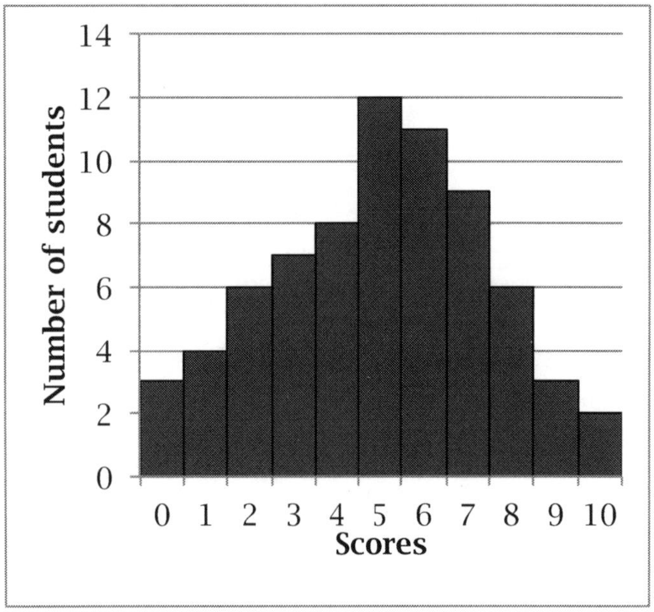

According to the histogram given above, what is the modal score (Mode) for 71 students?

(A) 4

(B) 5

(C) 6

(D) 12

154. SPR:

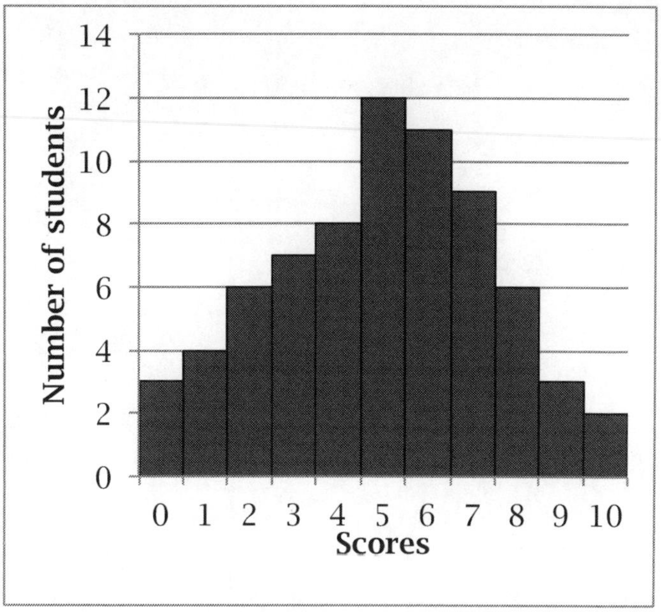

As per the data in the above histogram, what is the average (Mean) score of all 71 students?

Question on Range

155. Which of the following could be the range of set $\{-43, 76, a, b, c\}$?

I. -129

II. 112

III. 154

(A) I only

(B) II only

(C) III only

(D) Only I and II

Questions on Standard deviation

156. Which of the following sets have a standard deviation less than the standard deviation of set X: {10, 20, 30, 40}?

$P = \{2, 12, 22, 32\}$
$Q = \{-3, 7, 17, 27\}$
$R = \{-6, 4, 14, 24\}$

(A) Set P

(B) Set Q

(C) Set R

(D) None

157. If Set X is $\{p, q, r\}$, which of the following sets must have the same standard deviation as has Set X?

(A) $\{p + 1, q, r\}$

(B) $\{p, q + 1, r\}$

(C) $\{p, q, r + 1\}$

(D) $\{p + qr, q(1 + r), r(q + 1)\}$

158. If one of P and Q is a positive number and the other is a negative number, which of the following could be the standard deviation of set $\{P, -12, -8, -4, 0, 4, 8, Q\}$?

I. -4

II. -2

III. 0

(A) I only

(B) II only

(C) I and II only

(D) None of these

159. A group of five friends collected WWE stickers. They noticed that they had an average of 15 stickers per person. If the maximum number of stickers with any of them was 18 and each friend had a distinct number of stickers, what could be the minimum number of stickers with any of them?

(A) 3

(B) 6

(C) 9

(D) 12

3.2.10 Data Analysis

160. Percentage distribution of GDP & Government spending ($M) for Country U during year N is given by the following chart.

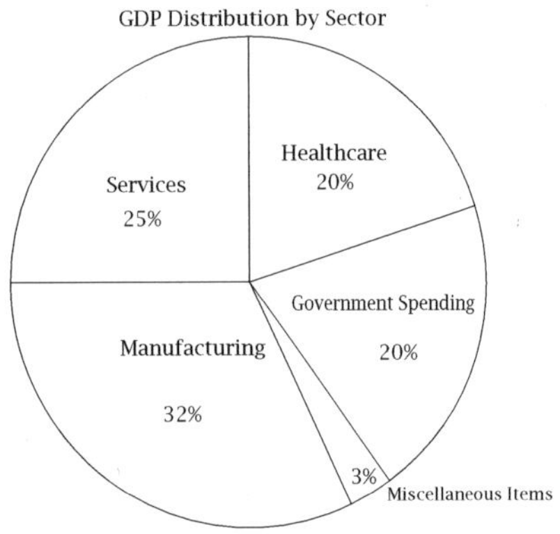

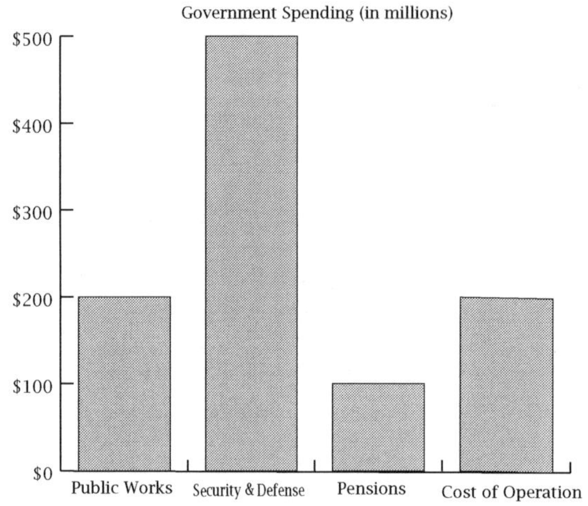

During Year N, the government's cost of operation was what percent of the total GDP of Country U?

(A) 4%

(B) 5%

(C) 8%

(D) 20%

161. Percentage distribution of GDP & Government spending ($M) for Country U during year N is given by the following chart.

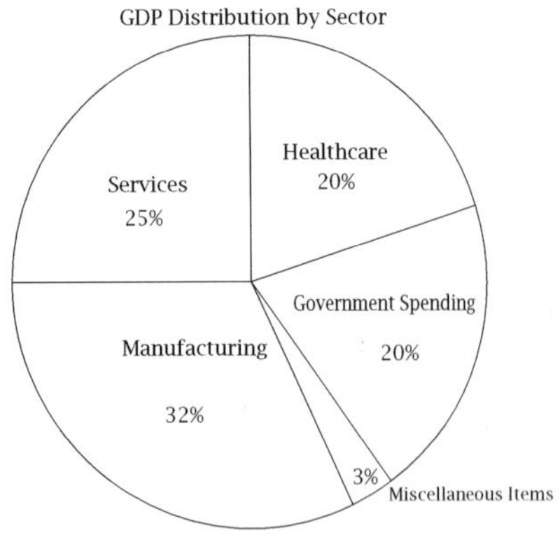

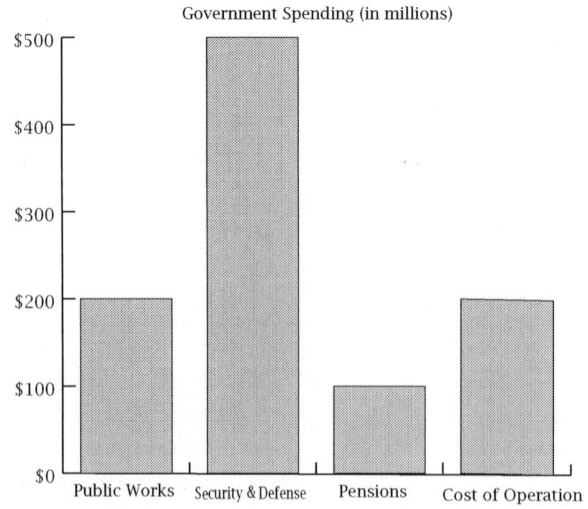

Making use of the information in the previous problem, if Management Consulting Services, a part of Services, in the economy of the country have the same dollar value as Miscellaneous Items, what percent of Services is accounted for by Management Consulting Services?

(A) 60%

(B) 15%

(C) 12%

(D) 6%

162.

Week	Number of data requests
1	20,000
2	20,000
3	15,000
4	5,000

The table above shows the approximate number of data requests at the internet site of a new reality show for the first 4 weeks after it went on air for the first time. The cost of advertisement per week on the show's website was computed to be 10% of the product of the number of data request per week and the average length of a data request per week. What was the total advertising cost for the first 4 weeks from the website, if the average length of a data request was estimated to be 4 minutes?

(A) $\$8,000$

(B) $\$18,000$

(C) $\$20,000$

(D) $\$24,000$

163.

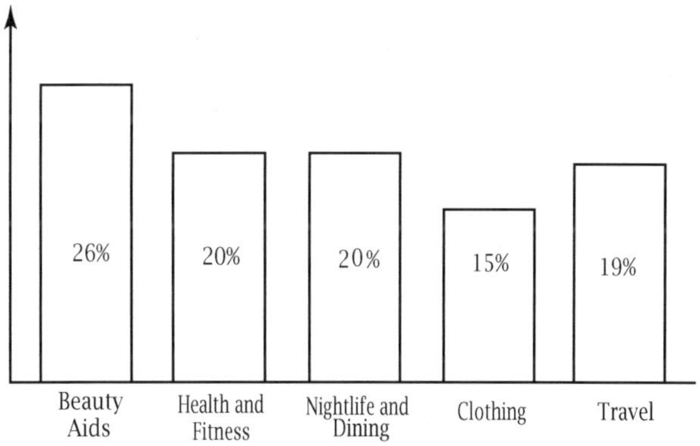

During a certain year Michelle spent her annual trust fund allowance according to the chart above. How much did Michelle spend on clothing, if she spent $38,000 on travel?

(A) $8,000

(B) $18,000

(C) $20,000

(D) $30,000

164. The following table presents quarter-wise sales and inventory for Apex Corporation for 10 steel products, in metric tons (MT).

S - Sales (MT); I - Inventory (MT)

Products	Quarter I		Quarter II		Quarter III		Quarter IV		Yearly Total	
	S	I	S	I	S	I	S	I	S	I
CR pipe	24	12	15	8	4	8	13	4	**56**	**32**
CR sheet	16	8	12	12	16	4	15	21	**59**	**45**
CR tube	15	15	16	8	6	15	8	4	**45**	**42**
CR wire	15	0	24	8	18	13	13	12	**70**	**33**
HR pipe	24	12	21	8	5	21	12	15	**62**	**56**
HR sheet	18	15	15	16	0	0	14	3	**47**	**34**
HR tube	8	9	8	21	8	0	9	2	**33**	**32**
HR wire	24	26	12	16	15	15	8	4	**59**	**61**
HS Billet	21	24	21	0	18	5	15	12	**75**	**41**
MS Billet	10	8	15	0	18	2	15	16	**58**	**26**
Total	**175**	**129**	**159**	**97**	**108**	**83**	**122**	**93**	**564**	**402**

Which of the following quarters has the highest average (arithmetic mean) sales in metric tons (MT)?

(A) Quarter I

(B) Quarter II

(C) Quarter III

(D) Quarter IV

165. The following table presents quarter-wise sales and inventory for Apex Corporation for 10 steel products, in metric tons (MT).

S - Sales (MT); I - Inventory (MT)

Products	Quarter I		Quarter II		Quarter III		Quarter IV		Yearly Total	
	S	I	S	I	S	I	S	I	S	I
CR pipe	24	12	15	8	4	8	13	4	**56**	**32**
CR sheet	16	8	12	12	16	4	15	21	**59**	**45**
CR tube	15	15	16	8	6	15	8	4	**45**	**42**
CR wire	15	0	24	8	18	13	13	12	**70**	**33**
HR pipe	24	12	21	8	5	21	12	15	**62**	**56**
HR sheet	18	15	15	16	0	0	14	3	**47**	**34**
HR tube	8	9	8	21	8	0	9	2	**33**	**32**
HR wire	24	26	12	16	15	15	8	4	**59**	**61**
HS Billet	21	24	21	0	18	5	15	12	**75**	**41**
MS Billet	10	8	15	0	18	2	15	16	**58**	**26**
Total	**175**	**129**	**159**	**97**	**108**	**83**	**122**	**93**	**564**	**402**

Which of the following statements is correct?

(A) In Quarter I, there is only one product that has the greatest value of sales.

(B) In Quarter II, for 'HS billet' and 'MS billet', the ratio of 'sale to inventory' is 0.

(C) In Quarter III, the product that has the highest inventory also has the least sale.

(D) In Quarter IV, the product that has the highest inventory also has the greatest sale.

166. The graph given below shows distribution of per unit cost ($) of an automotive component at different outputs of a machine. The best estimation curve is also shown with dashed line.

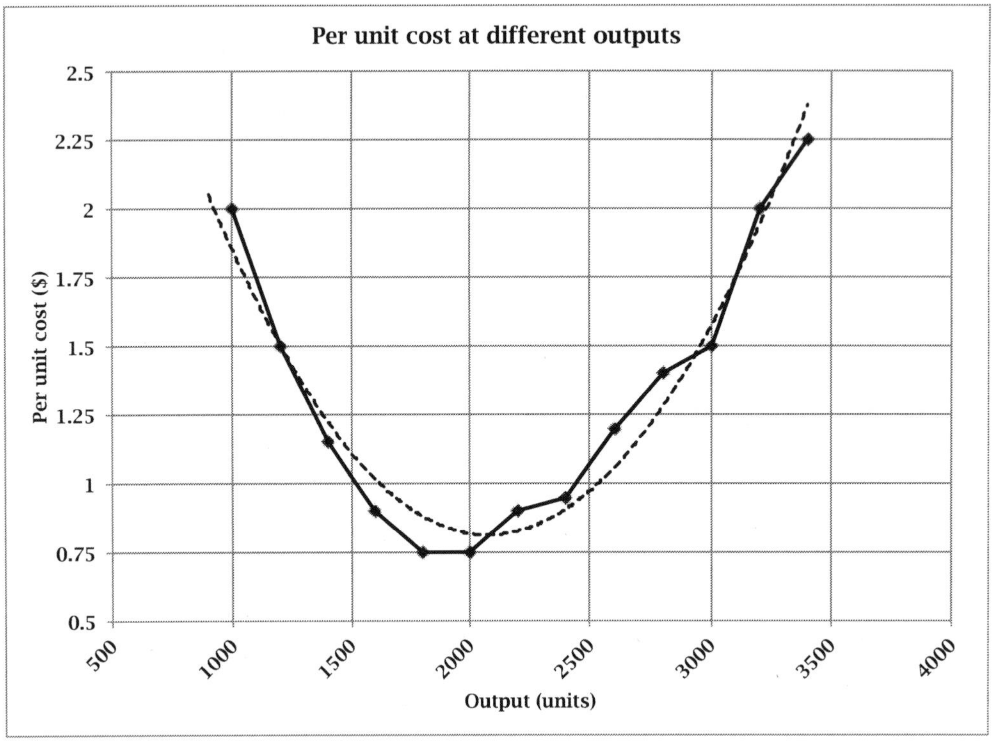

What is the best estimate for the output for the machine such that the per unit cost of the component is least?

(A) 800

(B) 1,000

(C) 2,100

(D) 3,400

167. The graph given below shows distribution of per unit cost ($) of an automotive component at different outputs of a machine. The best estimation curve is also shown with dashed line.

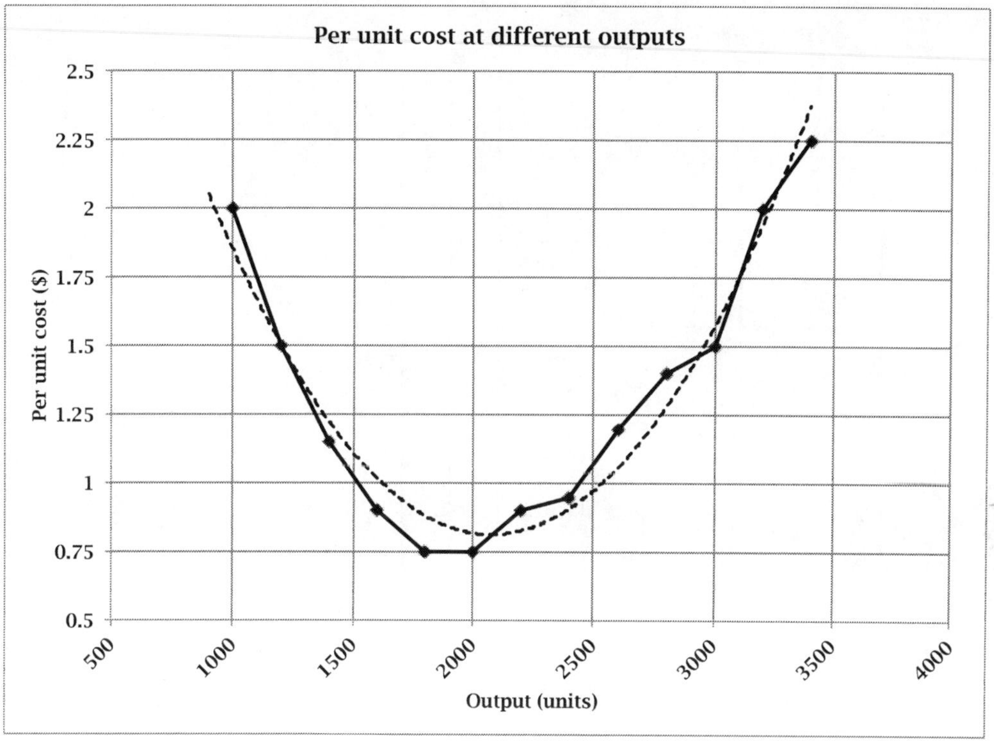

According to the above graph, at which of the following output would the per unit cost be the same as it is at 1,200 units of output?

(A) 1,000

(B) 2,000

(C) 3,000

(D) 3,400

168. The graph given below shows a scatterplot for the distribution of turnovers ($M) and their corresponding profits ($M). The data is modeled as a linear relationship (a dashed trend-line), where p is profit and t is turnover. The following two questions are based on it.

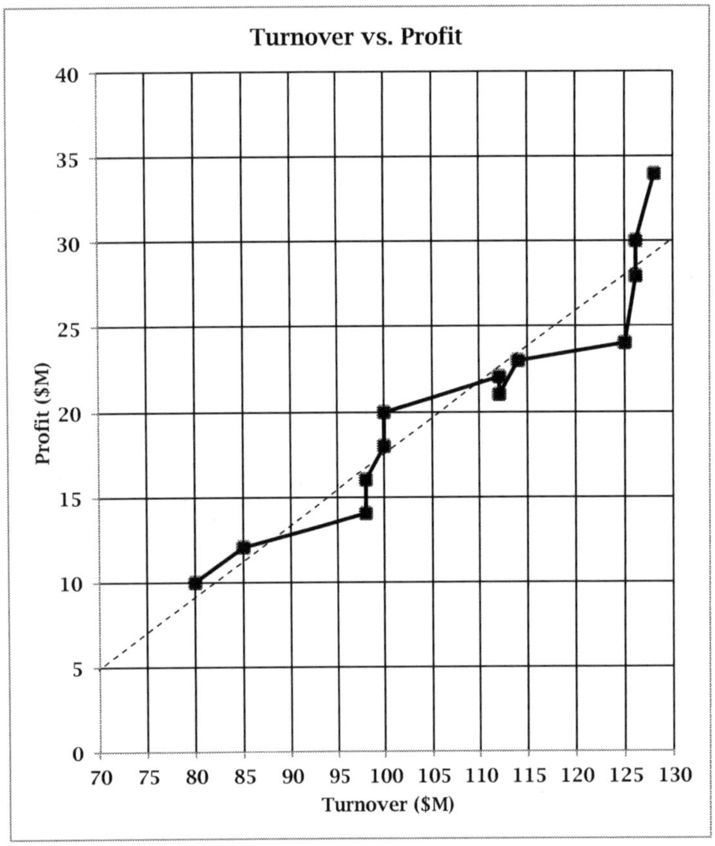

What is the estimated profit ($M) for the turnover of $125M?

(A) 25

(B) 28

(C) 30

(D) 125

169. The graph given below shows a scatterplot for the distribution of turnovers ($M) and their corresponding profits ($M). The data is modeled as a linear relationship (a dashed trend-line), where p is profit and t is turnover. The following two questions are based on it.

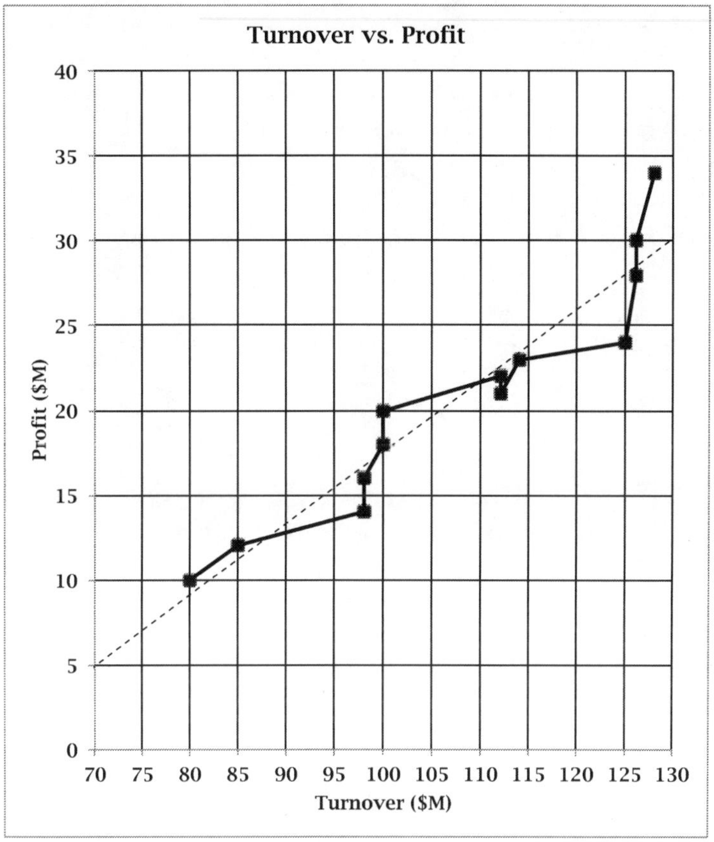

Deduce a linear equation that estimates the trend-line, using p and t?

(A) $p = 2.4t - 29.17$

(B) $p = 0.42t - 24.4$

(C) $p = 0.42t - 0.34$

(D) $p = 0.42t + 24.4$

170. The line graph given below shows numbers of complaints received for a product over the year for three plants. The following two questions are based on it.

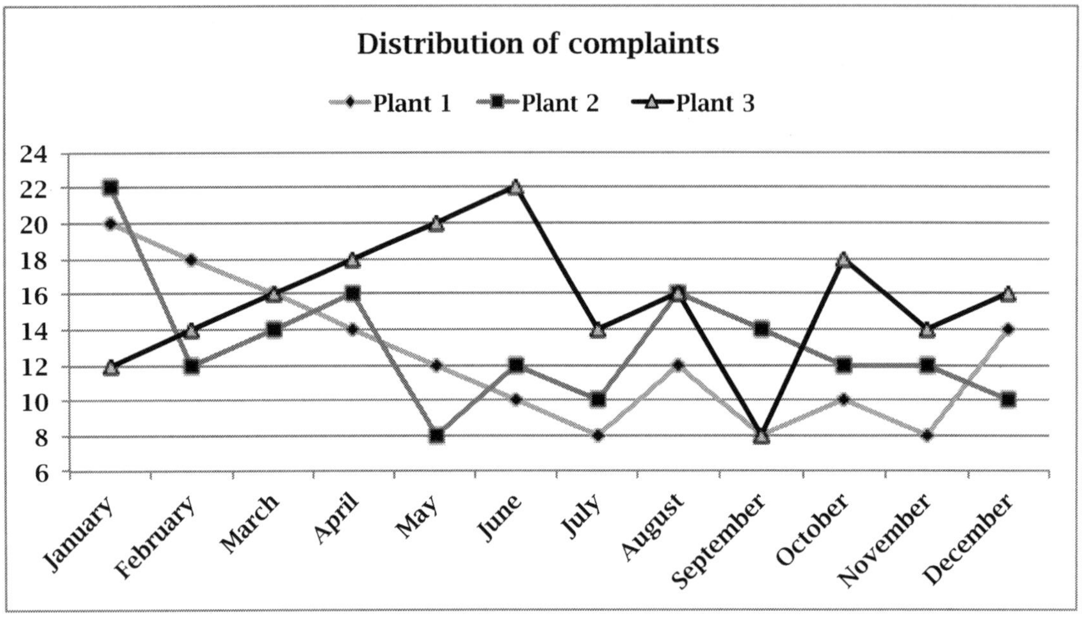

By what number is the number of complaints in three plants in the month of January is more than that in the month of June?

(A) 10

(B) 12

(C) 14

(D) 16

171. SPR:

The line graph given below shows numbers of complaints received for a product over the year for three plants. The following two questions are based on it.

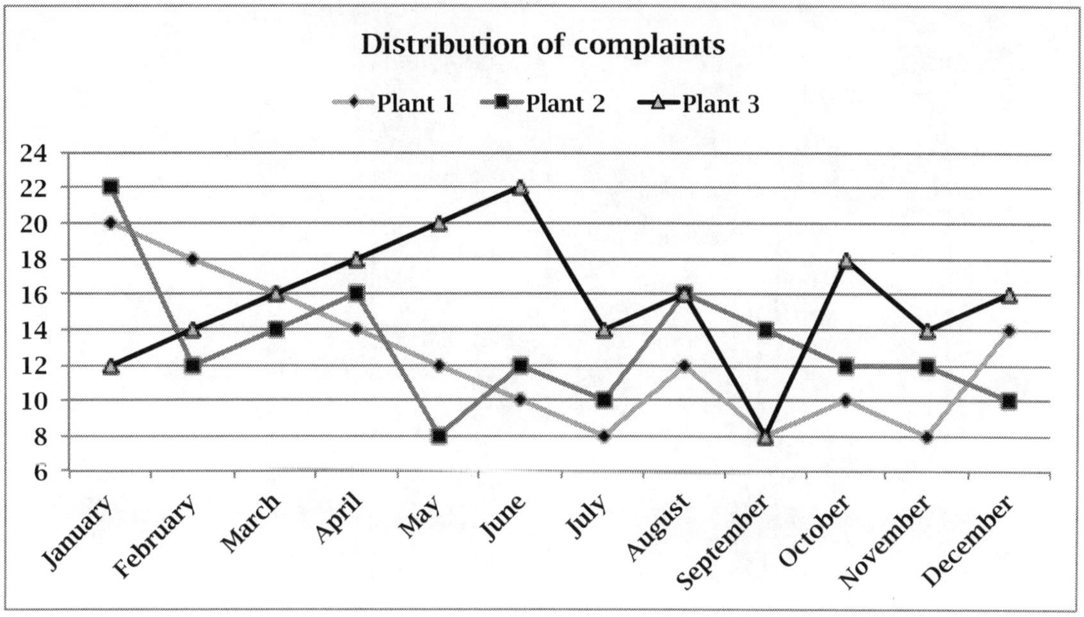

For how many corresponding months is the number of complaints for Plant 3 higher than the number of complaints for both Plant 1 and for Plant 2?

172. The bar chart given below shows the percentage distribution of expenses for following heads: Food, Clothing, Home, and Entertainment for three families. Total expenses incurred by each family are also mentioned in the chart.

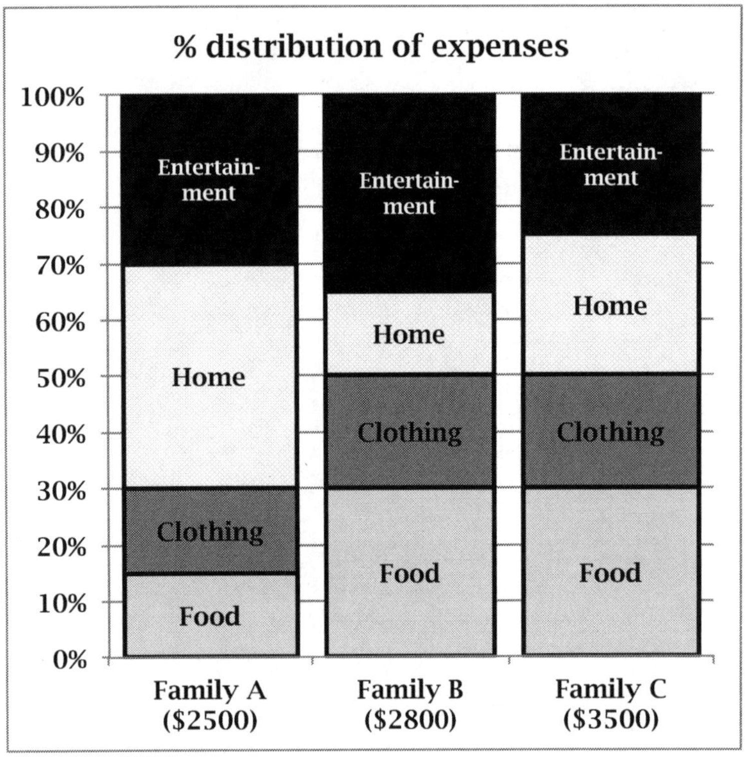

Which of the following families spent highest on entertainment?

(A) Family A

(B) Family B

(C) Family C

(D) Families A & B

173. SPR:

The bar chart given below shows the percentage distribution of expenses for following heads: Food, Clothing, Home, and Entertainment for three families. Total expenses incurred by each family are also mentioned in the chart.

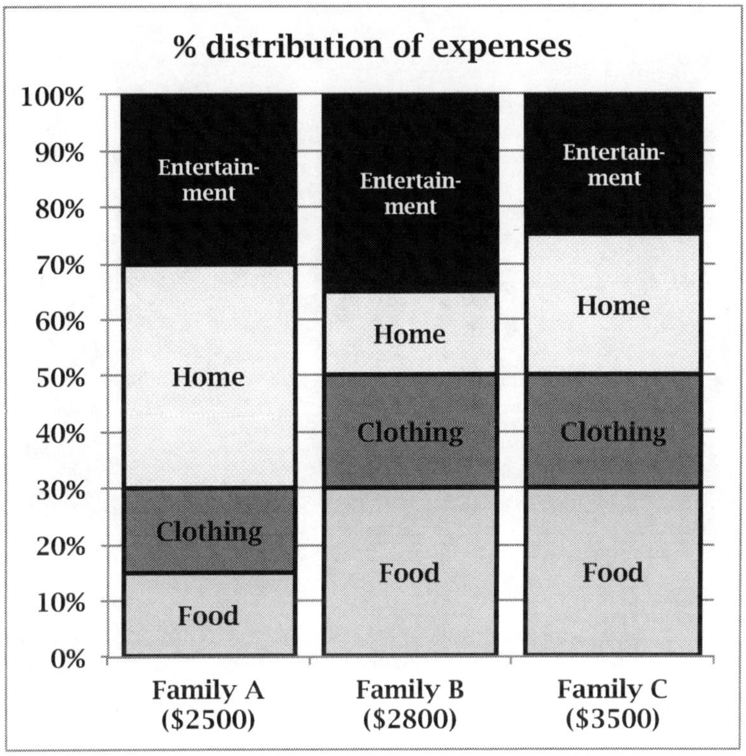

Of the families, how many dollars does the family that spends the least on 'Clothing' spend on 'Clothing'?

174. The multiple bar chart given below shows distribution of commuters for 5 years on the basis of mode of transport used by them, namely, Intra-City Bus, Local Train, and Private taxi.

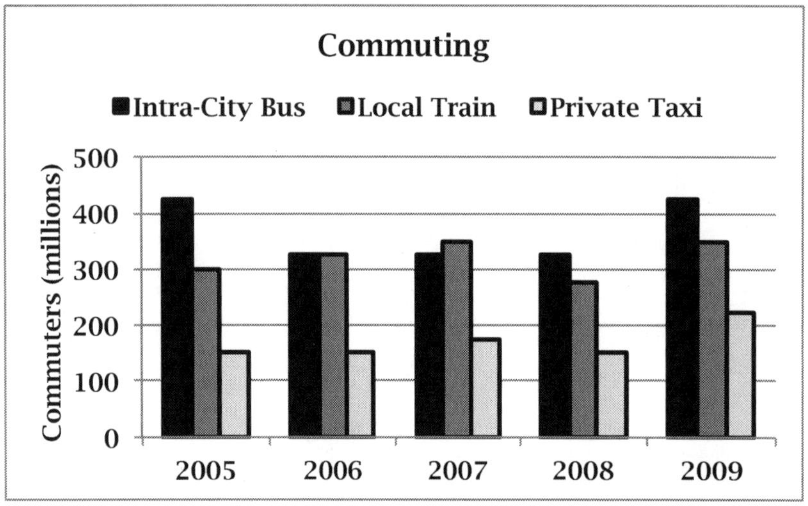

In which of the following years, did the maximum number of commuters travel?

(A) 2006

(B) 2007

(C) 2008

(D) 2009

175. The multiple bar chart given below shows distribution of commuters for 5 years on the basis of mode of transport used by them, namely, Intra-City Bus, Local Train, and Private taxi.

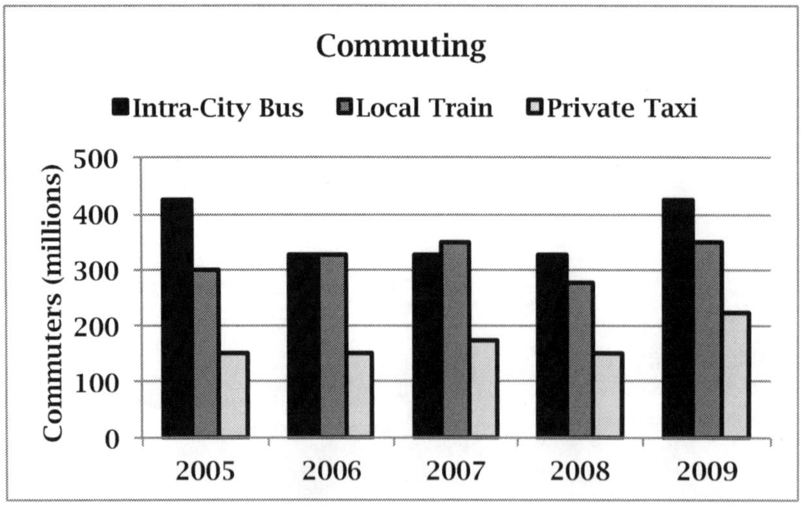

Of the five years, approximately how many commuters traveled (in millions) in the year in which the least number of commuters traveled?

(A) 600

(B) 650

(C) 750

(D) 850

3.3 Advanced Math

3.3.1 Polynomials

176. Which of the following expressions is equivalent to $(2x^5 - 3x^4 - 4x^2 + 3) - (2x^4 - 3x^2 + 4x + 13) - (2x^3 - 3x^2 - 4x)$?

(A) $2x^5 - 3x^4 + 8x + 16$

(B) $2x^5 - 5x^4 + 2x^2 - 10$

(C) $2x^5 - x^4 + 2x^2 + 8x - 10$

(D) $2x^5 - 3x^4 + 2x^2 + 16$

177. Which of the following expressions is the equivalent to $13x^5y^{-3}z^2 \times 4x^{-2}y^3z^{-3}$

(A) $52x^{-10}y^{-9}z^{-6}$

(B) $52x^3z^{-1}$

(C) $52x^3yz^{-6}$

(D) $52x^{-10}y^{-1}z^6$

178. Which of the following expressions is equivalent to $(-3x^4 + 3) \times (2x^4 + 4x)$?

(A) $-6x^8 + 3x^5 + 6x^4 + 12x$

(B) $-6x^8 - 12x^5 + 6x^4 + 12x$

(C) $-6x^8 + 12x^5 - 6x^4 + 12x$

(D) $-6x^{16} - 3x^5 + 6x^4 + 12x$

179. Which of the following expressions represents $\frac{6x^5-33x^4+21x^3-6x^2-30x}{3x}$?

(A) $2x^4 - 11x^3 + 7x^2 - 2x - 10x$

(B) $2x^4 - 11x^3 + 7x^2 - 2x - 10$

(C) $2x^4 - 11x^4 + 7x^2 - 2x - 10$

(D) $2x^4 - 11x^3 + 7x^2 - 2x$

180. Which of the following expressions represents $\frac{3x^3-24x^2-4x+32}{x-8}$?

(A) $x - 8$

(B) $3x^2 - 2x - 4$

(C) $3x^2 - 4$

(D) $(x - 8)(3x^2 - 4)$

181. Which of the following expressions represents $\frac{8x^3+125}{2x+5}$?

(A) $4x^2 - 10x - 25$

(B) $4x^2 - 20x + 25$

(C) $4x^2 - 10x + 25$

(D) $4x^2 + 10x + 25$

182. Which of the following expressions represents $\frac{8-729x^6}{\sqrt{2}-3x}$?

(A) $(2 + 3x)(4 + 18x^2 + 81x^4)$

(B) $(\sqrt{2} - 3x)(4 + 18x^2 + 81x^4)$

(C) $4 + 18x^2 + 81x^4$

(D) $(\sqrt{2} + 3x)(4 + 18x^2 + 81x^4)$

3.3.2 Quadratic equations

183. **SPR:**

What is the integer value of x that satisfies the equation $\frac{3}{x-1} - \frac{2}{2x-3} = 1$?

184. What is the sum of values of x that satisfies the equation $\frac{3}{x} + \frac{x}{6} = \frac{3}{2}$?

(A) 2

(B) 3

(C) 6

(D) 9

185. What is the integer value of x that satisfies the equation $\frac{12}{x+4} + \frac{6}{x-6} = 4$?

(A) -3

(B) -8

(C) 8

(D) 12

186. If $x^2 - 3x + 2 = 0$ and $y^2 - 8y + 15 = 0$, which of the following is correct?

(A) Both values of x are greater than both values of y.

(B) Both values of y are greater than both values of x.

(C) Both values of x are equal to both values of y.

(D) One value of the values of y is greater than one value of of the values of x.

Graphic polynomial and quadratic equations

187.

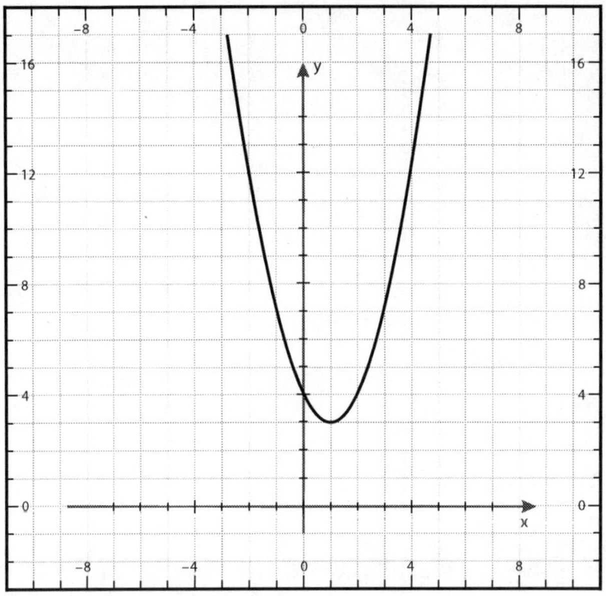

The graph of $f(x) = (x-1)^2 + 3$ is shown above. Choose from the options below the correct graph of $g(x) = (x+1)^2 + 1$.

(A)

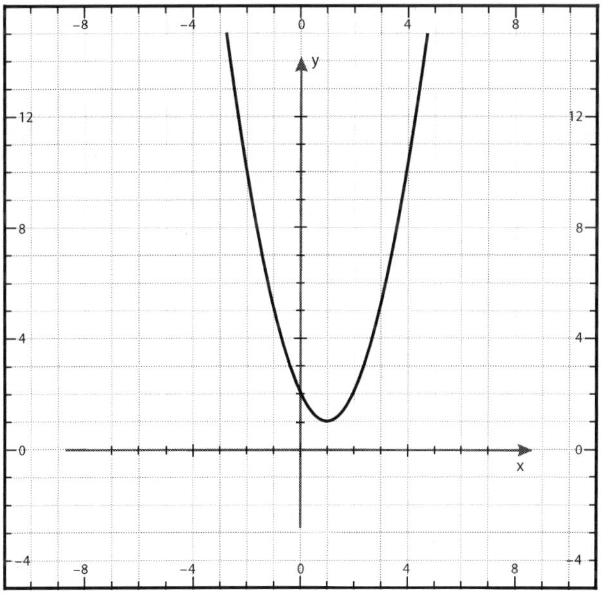

(B)

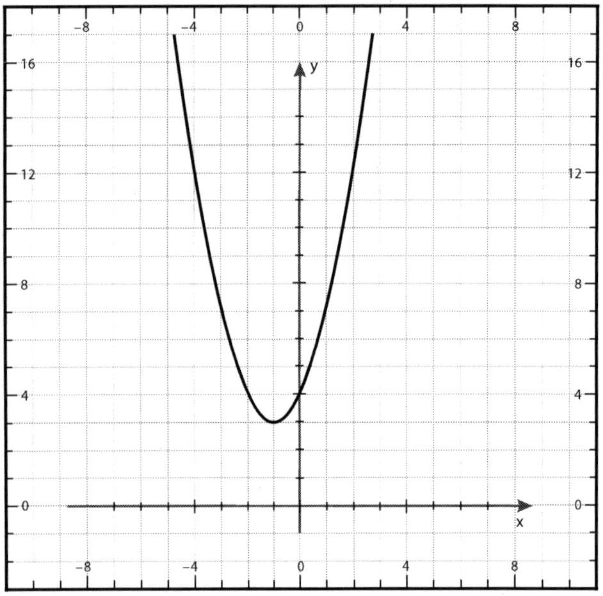

(C)

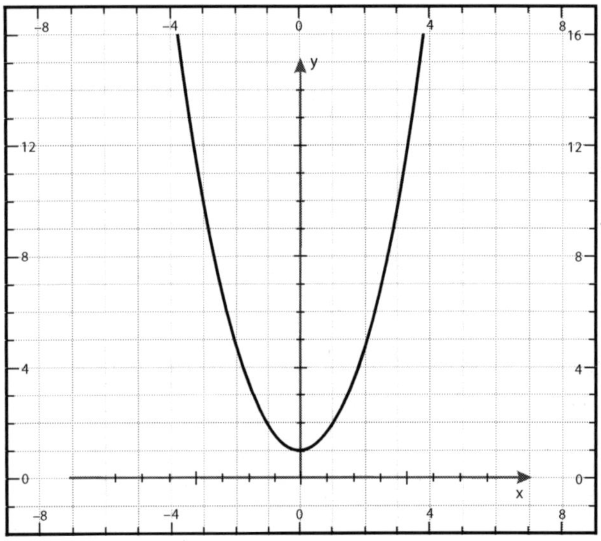

(D)

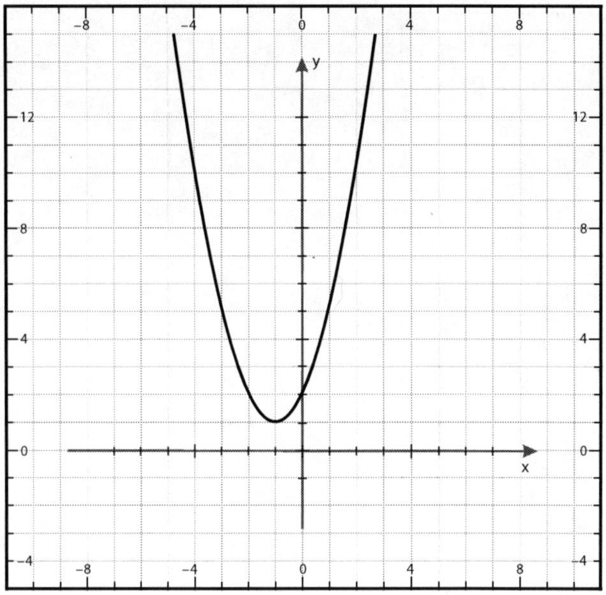

188.

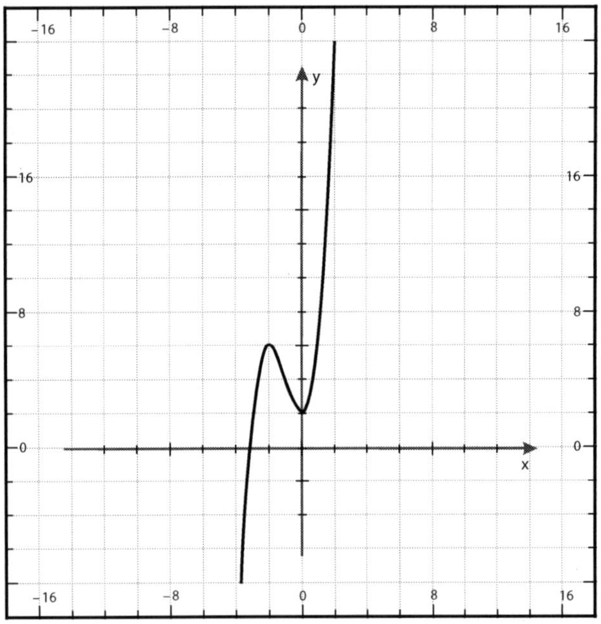

$y = f(x)$

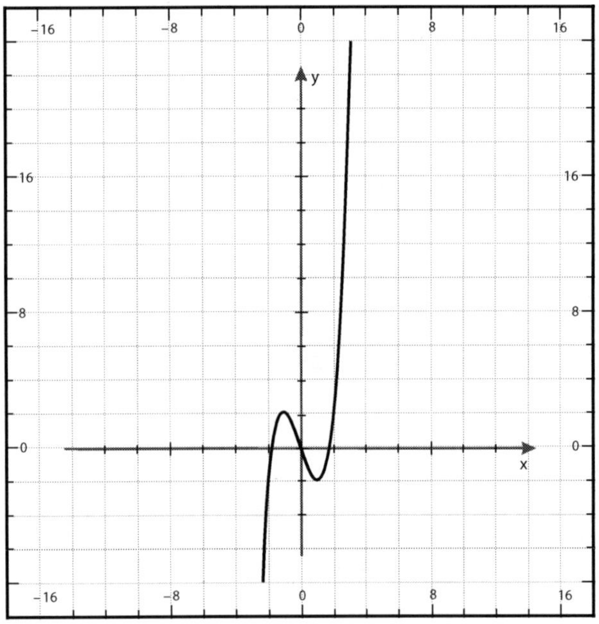

$y = g(x)$

The graphs of $f(x)$ and $g(x)$ are shown above. Choose from the options below the correct relation between $g(x)$ and $f(x)$.

(A) $g(x) = f(x+1)$

(B) $g(x) = f(x+1) - 2$

(C) $g(x) = f(x-1) + 4$

(D) $g(x) = f(x-1) - 4$

189.

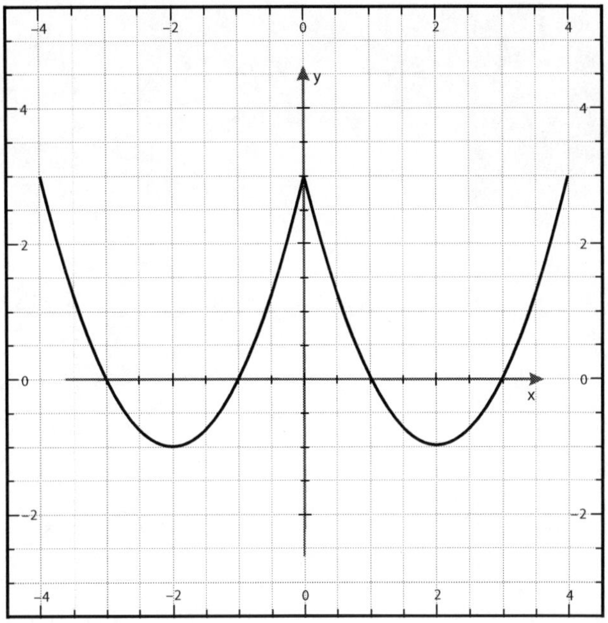

$y = f(x)$

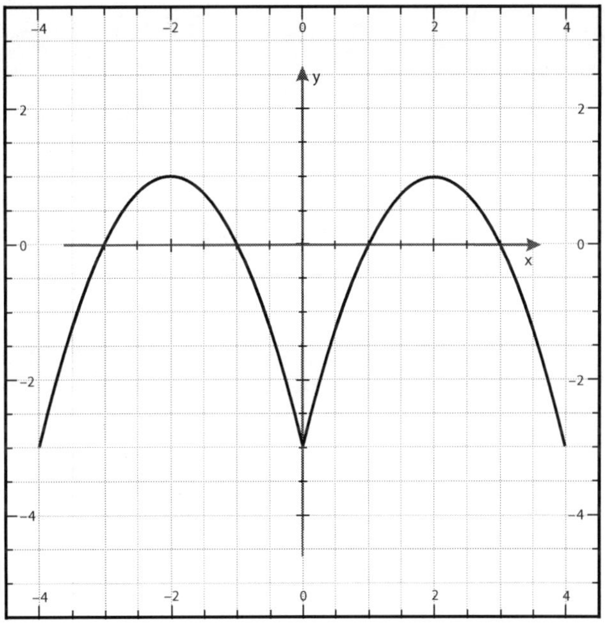

$y = g(x)$

The graphs of $f(x)$ and $g(x)$ are shown above. Choose from the options below the correct relation between $g(x)$ and $f(x)$.

(A) $g(x) = f(x - 1)$

(B) $g(x) = f(-x)$

(C) $g(x) = -f(x)$

(D) $g(x) = f(x) - 12$

190.

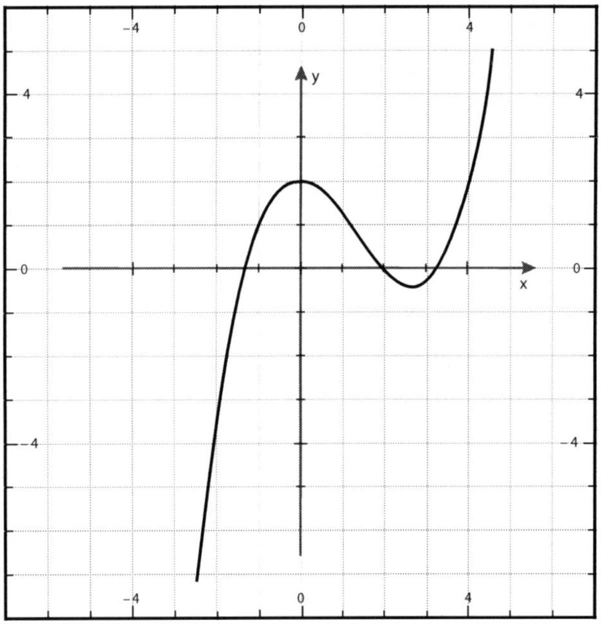

$y = f(x)$

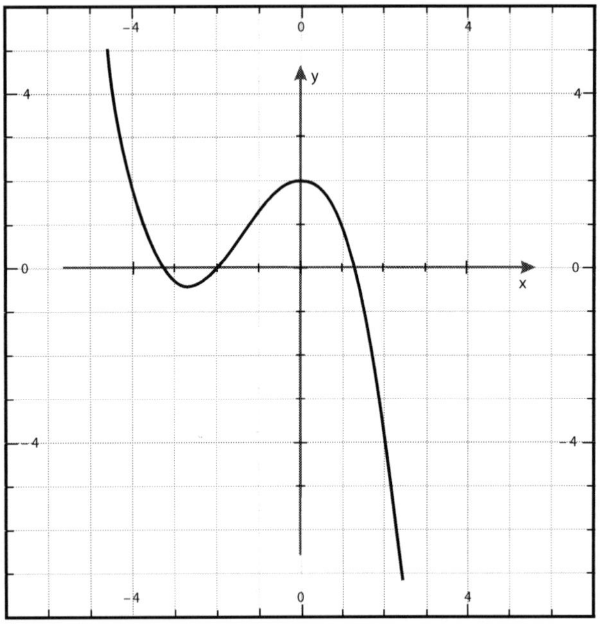

$y = g(x)$

The graphs of $f(x)$ and $g(x)$ are shown above. Choose from the options below the correct relation between $g(x)$ and $f(x)$.

(A) $g(x) = f(x) + 4$

(B) $g(x) = f(-x)$

(C) $g(x) = -f(x)$

(D) $g(x) = f(x + 4)$

191.

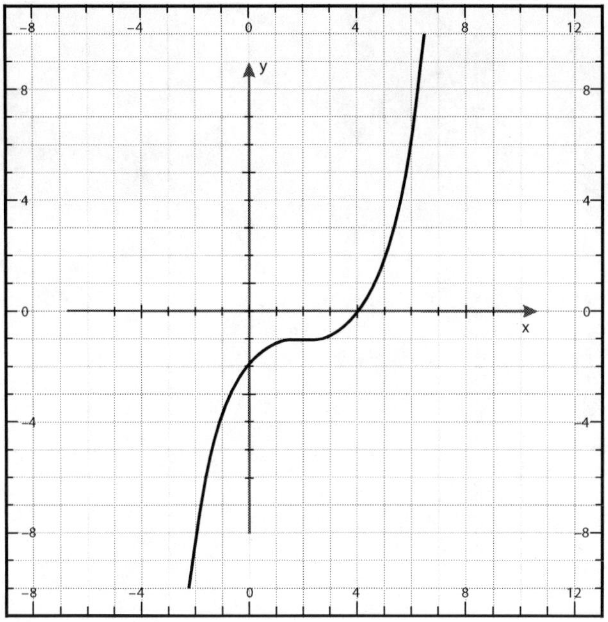

$y = f(x)$

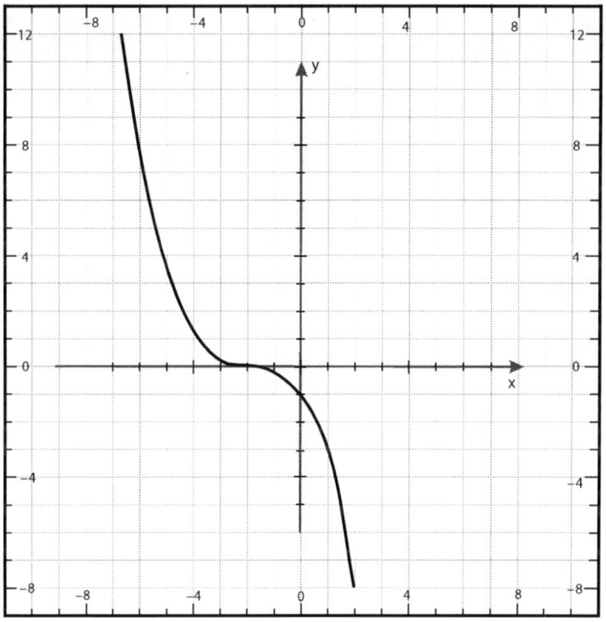

$y = g(x)$

The graphs of $f(x)$ and $g(x)$ are shown above. Choose from the options below the correct relation between $g(x)$ and $f(x)$.

(A) $g(x) = -f(x) + 1$

(B) $g(x) = f(-x) + 1$

(C) $g(x) = -f(x) - 1$

(D) $g(x) = f(-x) - 1$

192.

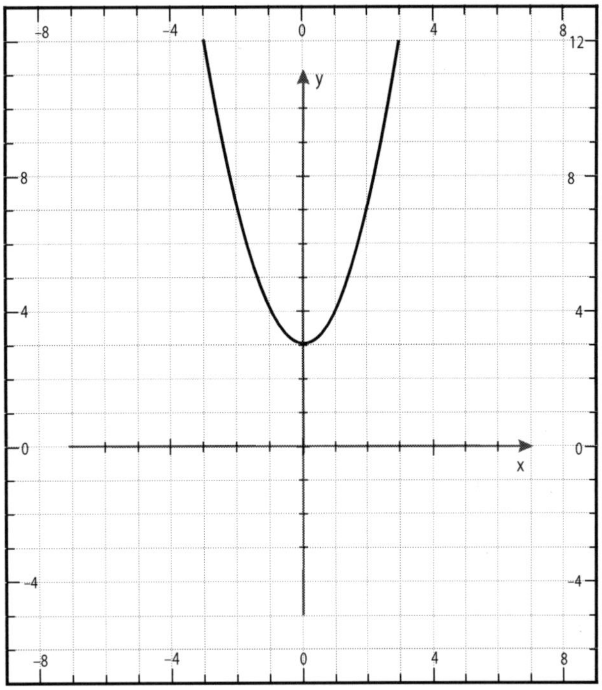

$f(x) = x^2 + 3$

The graph of $f(x) = x^2 + 3$ is shown above. If $g(x) = \frac{x^2}{3} + 3$, then what changes, if made to the graph of $f(x)$ would make it look similar to the graph of $g(x)$?

(A) Shift the graph of $f(x)$ up by 2 units

(B) Shift the graph of $f(x)$ down by 2 units

(C) Make the graph of $f(x)$ narrow but do not shift the graph

(D) Make the graph of $f(x)$ wide but do not shift the graph

193.

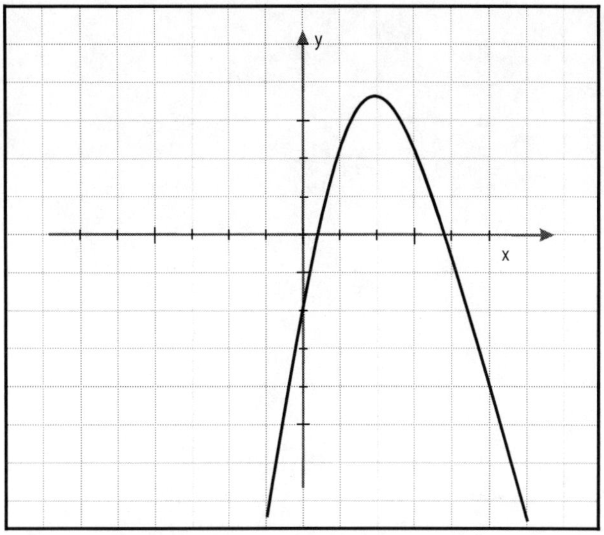

The graph of the quadratic function $f(x) = ax^2 + bx + c$ is shown above. Choose the correct statement from the options below.

(A) $a > 0, b < 0, c > 0$

(B) $a > 0, b > 0, c > 0$

(C) $a > 0, b > 0, c < 0$

(D) $a < 0, b > 0, c < 0$

194.

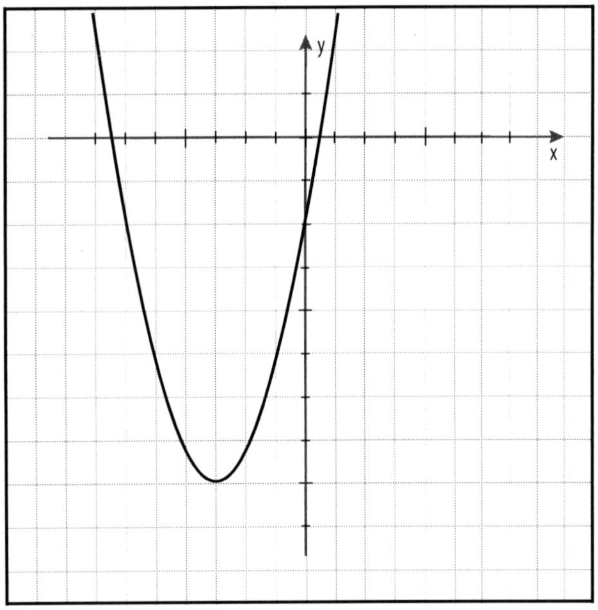

The graph of the quadratic function $f(x) = ax^2 + bx + c$ is shown above. Choose the correct statement from the options below.

(A) $a > 0, b < 0, c > 0$

(B) $a > 0, b > 0, c > 0$

(C) $a > 0, b > 0, c < 0$

(D) $a < 0, b > 0, c < 0$

195.

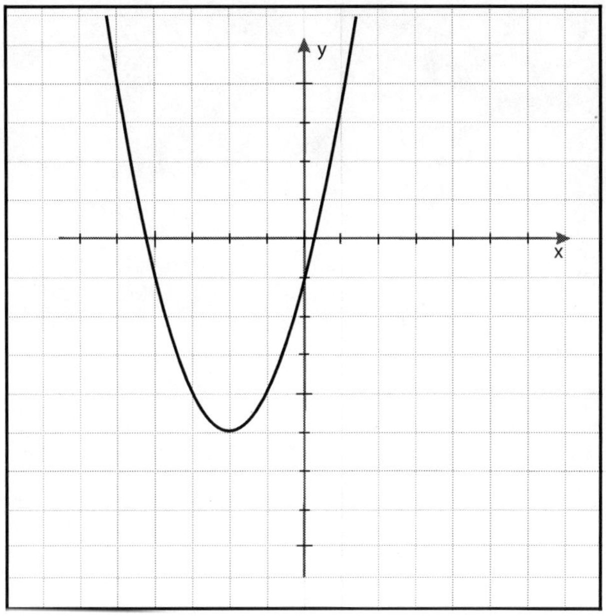

The graph of the quadratic function $f(x) = ax^2 + bx + c$ is shown above. Choose the correct statement from the options below for $f(x)$.

(A) The intercept on the y axis is equal to a

(B) The intercept on the y axis is equal to $-|b|$

(C) The intercept on the y axis is equal to $-b$

(D) The intercept on the y axis is equal to $-|c|$

196. Which of the following could be the correct graph of $f(x) = x^2 - 12x + 20$?

(A)

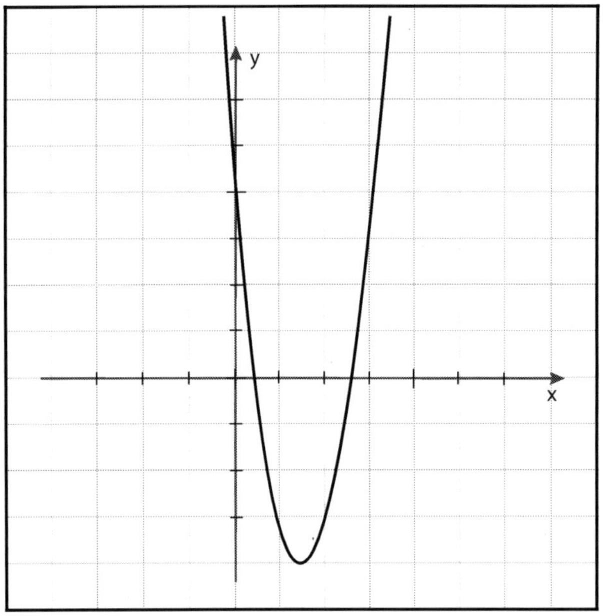

(B)

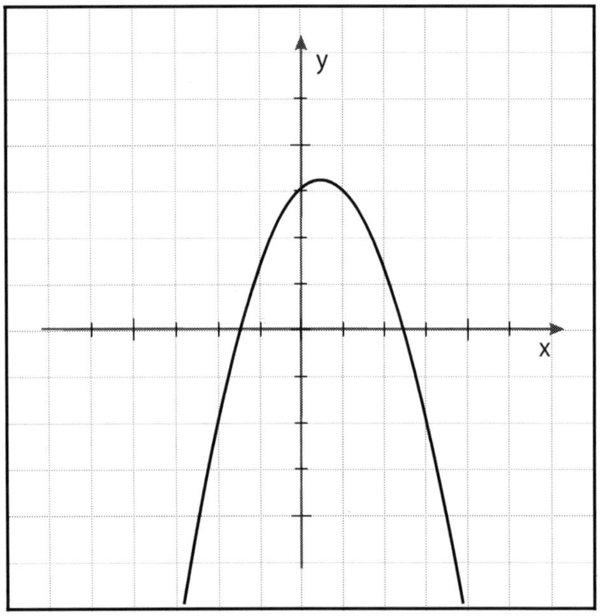

(C)

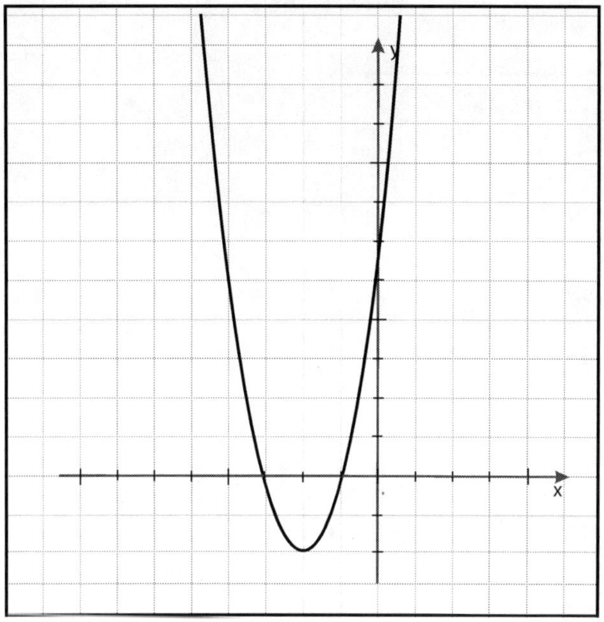

(D)

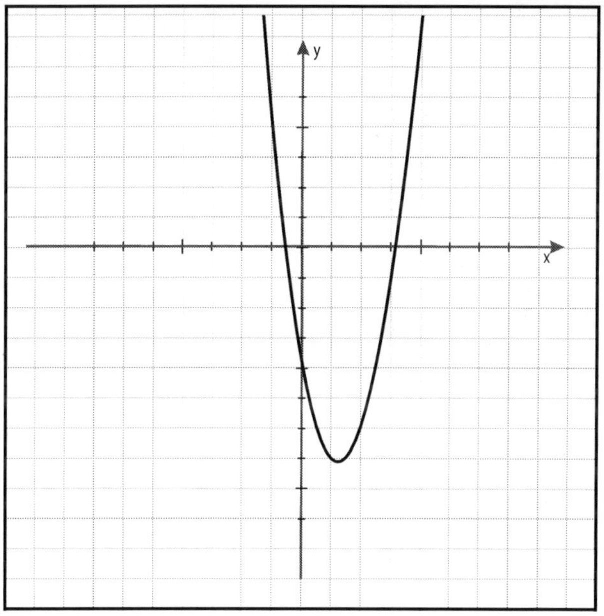

197. Which of the following could be the correct graph of $f(x) = x^2 + 6x + 9$?

(A)

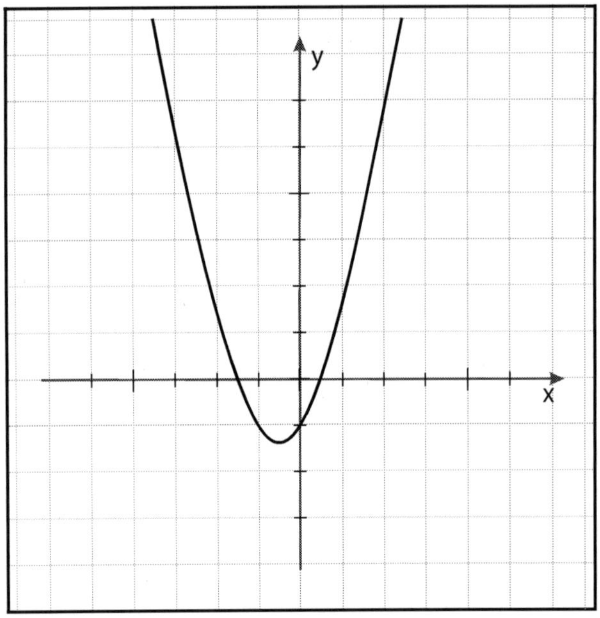

(B)

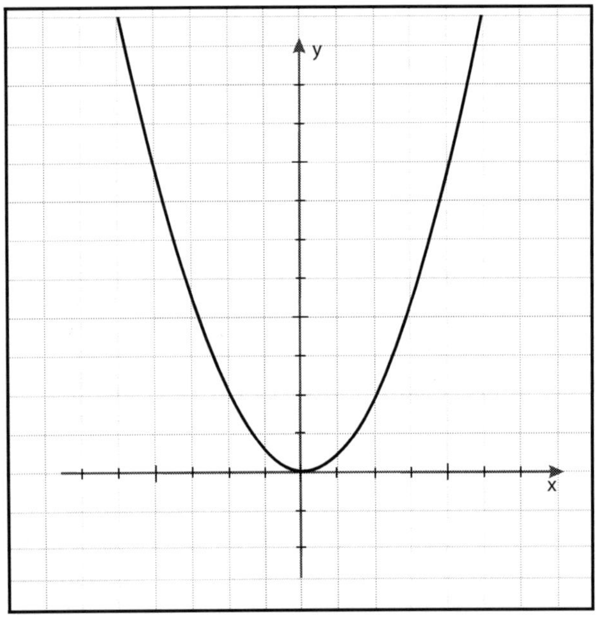

(C)

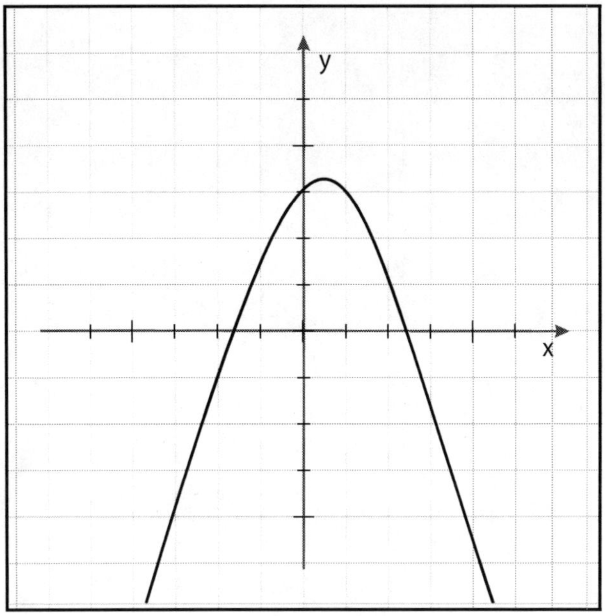

(D)

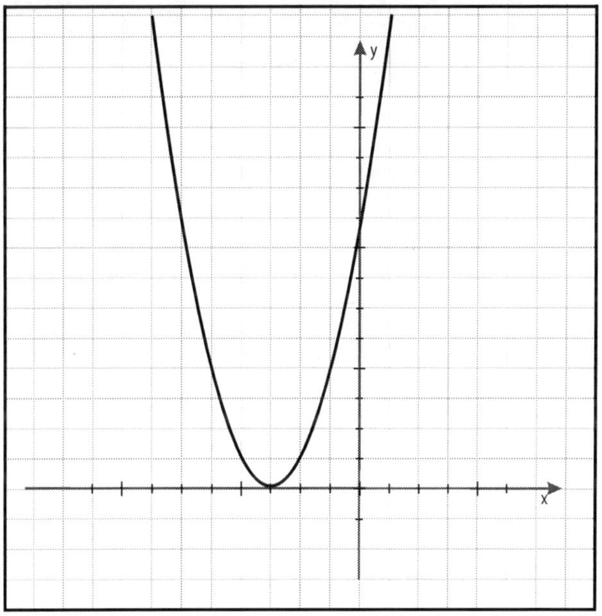

198.

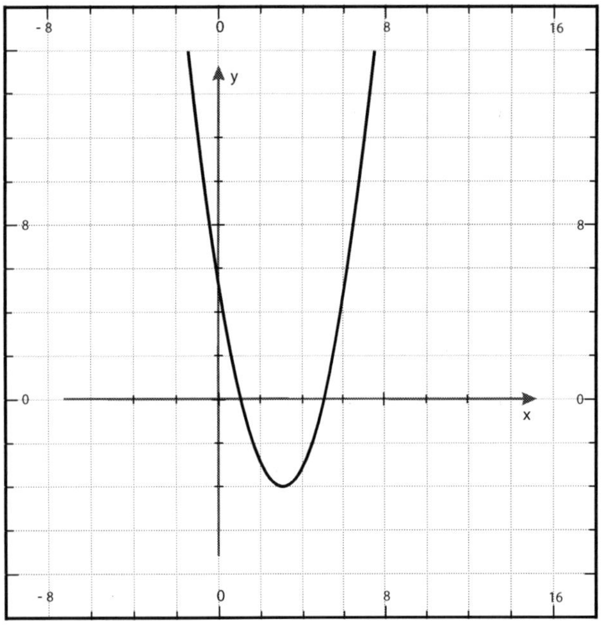

$y = f(x)$

Which one is the correct expression that corresponds to the graph of $f(x)$ shown above?

(A) $f(x) = x^2 + 8$

(B) $f(x) = x^2 + 4x + 6$

(C) $f(x) = (x - 3)^2 - 4$

(D) $f(x) = -x^2 + 4x - 8$

199. Which of the following could be the correct graph of $f(x) = x^2 + (a^2 - b)x + b^2 - a^2$, given that a and b are positive with $a > 1$ and $0 < b < 1$?

(A)

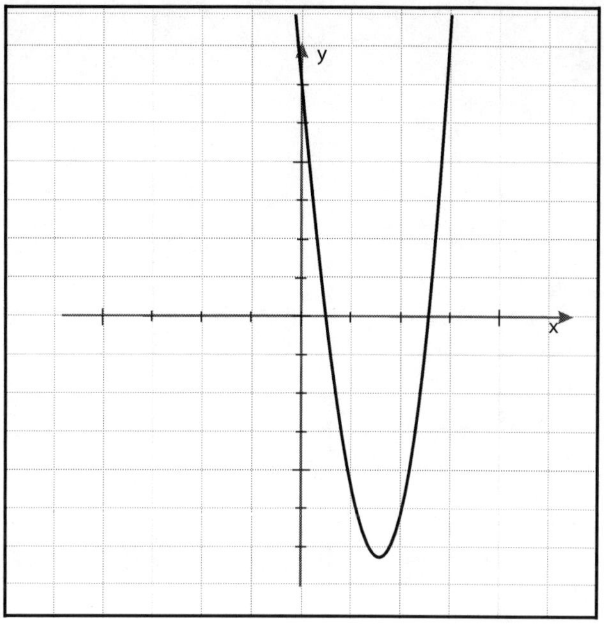

(B)

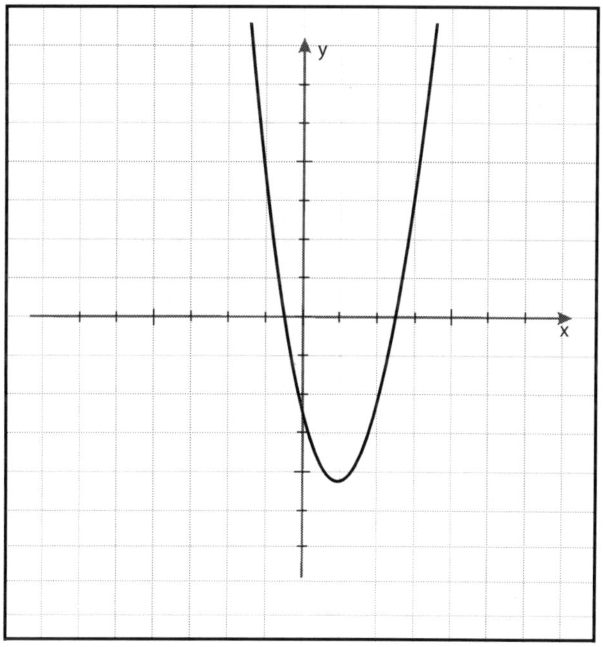

(C)

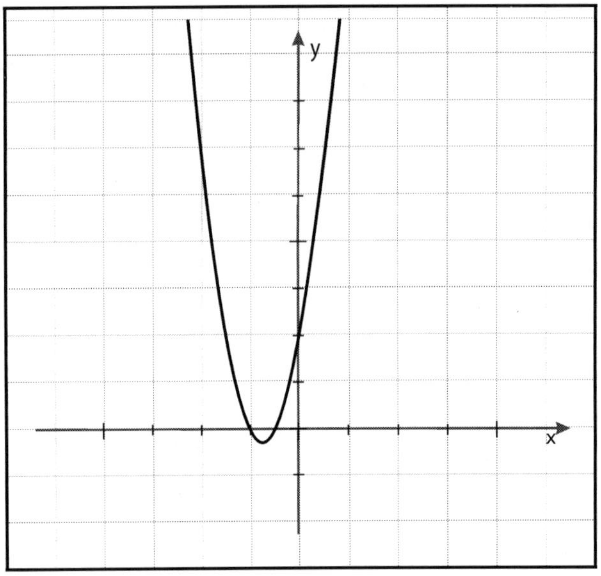

(D)

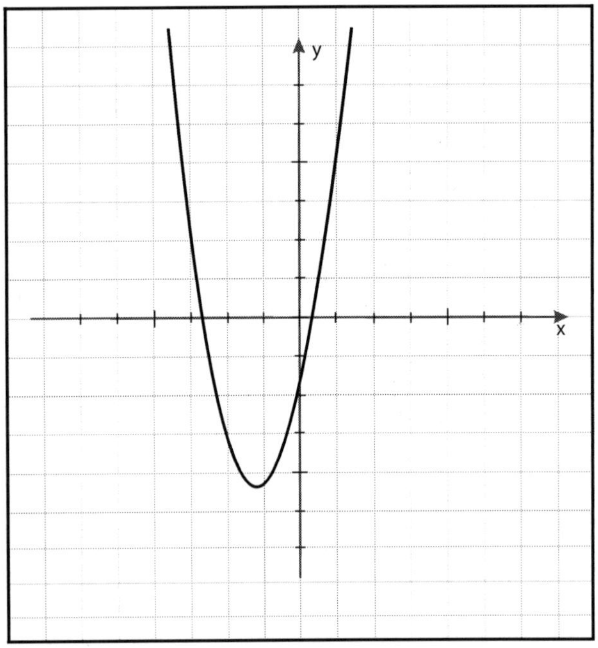

200. Which of the following is the correct graph of $f(x) = -x^2 + (a - b)x + b^2$, given that a and b are positive with $0 < a < 1$ and $3 < b < 5$?

(A)

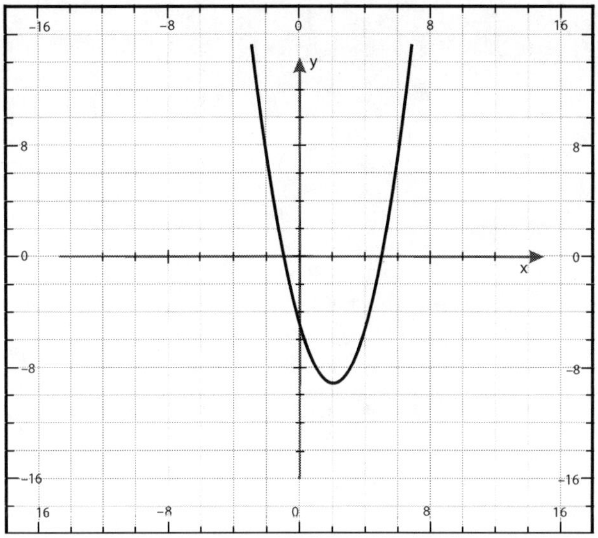

(B)

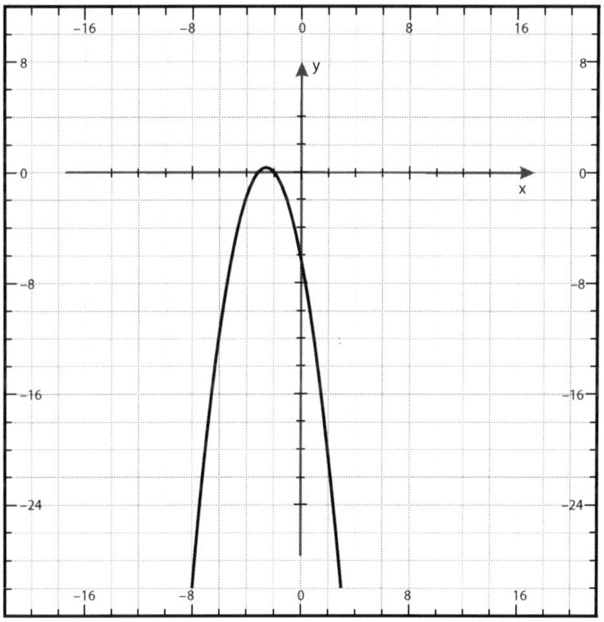

(C)

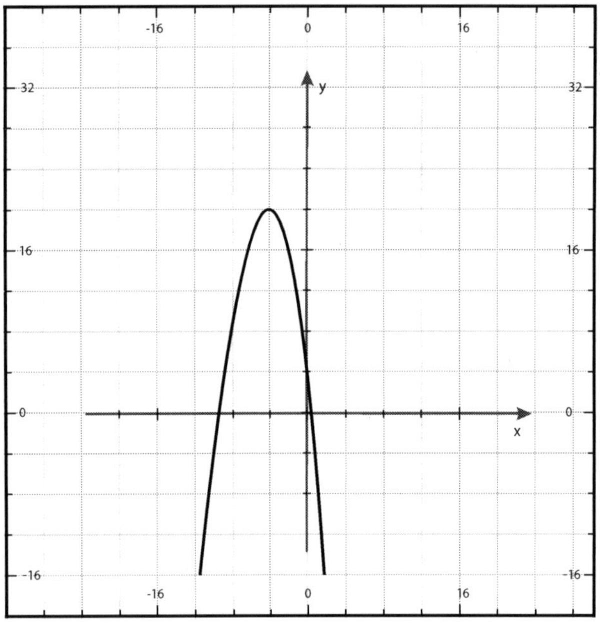

(D)

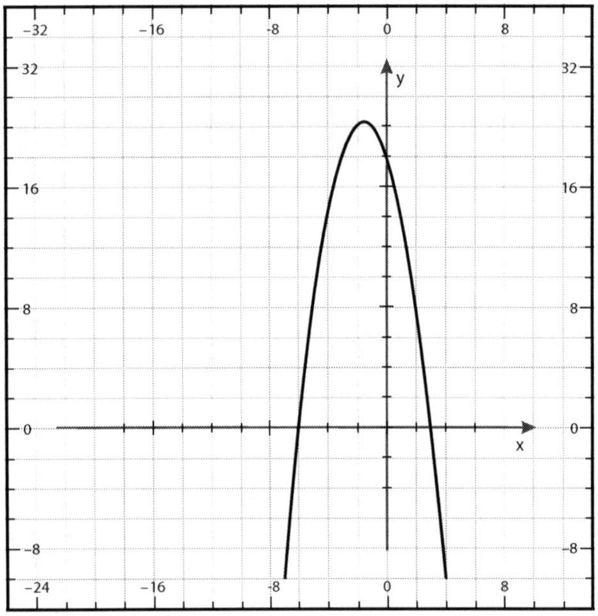

201. SPR:

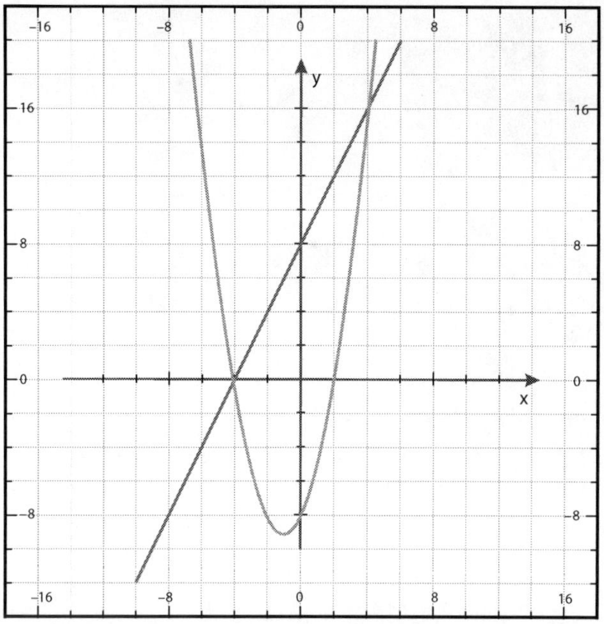

The graphs of $f(x) = x^2 + 2x - 8$ and $g(x) = 2x + 8$ are shown above. For how many integer values of x is $g(x) > f(x)$?

202. SPR:

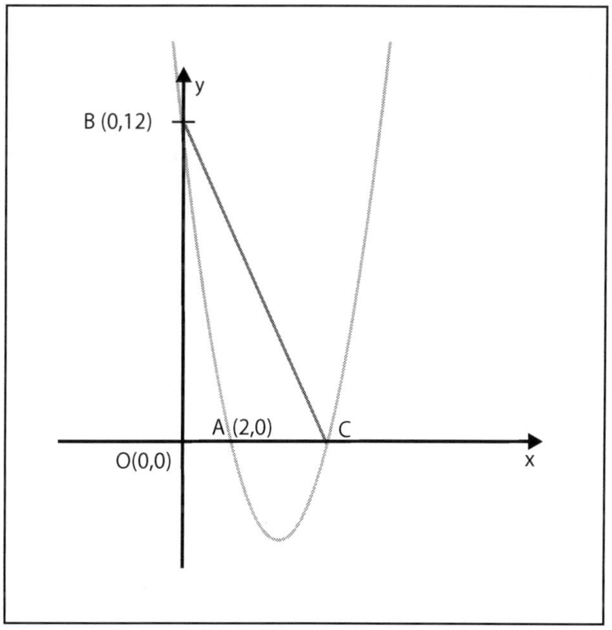

The graph of $f(x) = x^2 + ax + b$ is shown above. What is the area of the $\triangle$BOC?

3.3.3 Advanced equations

203. What is the value of x that satisfies the equation $\sqrt{2x-8} = \sqrt{x+5} - 3$?

(A) 4

(B) 12

(C) 36

(D) 76

204. What is the non-zero value of x that satisfies the equation: $\left(2^{x+2}\right)^{x+6} = 4^6$?

(A) -8

(B) -2

(C) 2

(D) 8

205. **SPR:**
What is the value of x that satisfies the equation $3^{2x-1} = 243$?

206. What is the absolute difference between the values of x that satisfy the equation $\sqrt{3x-8} - x + 2 = 0$?

(A) -1

(B) 1

(C) 3

(D) 4

207. How many values of x satisfy the equation $\left(x^2 - 17\right)^2 = 64$?

(A) One

(B) Two

(C) Three

(D) Four

208. **SPR:**

What is the value of x that satisfies the equation $\sqrt{x-8} = \sqrt{x+8} - 2$?

209. What is the integer value of x that satisfies the equation $\sqrt{(-2)^{x-5}} = 2^{(x-6)}$?

(A) 4

(B) 7

(C) 8

(D) 10

3.3.4 Application of functions

210. Pete fires a pistol pointing it upwards. The height, in feet, attained by the bullet above the ground t seconds after the pistol was fired is given by $h(t) = -t^2 + 12t + 30$.

What is the maximum height above the ground level reached by the bullet?

(A) 30 ft.

(B) 36 ft.

(C) 60 ft.

(D) 66 ft.

211. Pete fires a pistol pointing it upwards. The height, in feet, attained by the bullet above the ground t seconds after the pistol was fired is given by $h(t) = -t^2 + 12t + 30$.

What is the earliest time when the height of the bullet above the ground is 57 feet?

(A) 2 seconds

(B) 3 seconds

(C) 6 seconds

(D) 9 seconds

212. Pete fires a pistol pointing it upwards. The height, in feet, attained by the bullet above the ground t seconds after the pistol was fired is given by $h(t) = -t^2 + 12t + 30$.

By what time does the bullet reach the same level above the ground as the level from which the bullet was fired?

(A) 6 seconds

(B) 9 seconds

(C) 12 seconds

(D) 18 seconds

213. The total cost, in dollars, of manufacturing flower vases by the Potter House follows the function $c(x) = x^2 - 17x + 80$, where x denotes the number of vases manufactured. The vases are sold at a fixed price of $25 per vase.

For how many of the following vases manufactured by the company will there be neither profit nor loss?

(A) 3

(B) 40

(C) 72

(D) 123

214. The total cost, in dollars, of manufacturing flower vases by the Potter House follows the function $c(x) = x^2 - 17x + 80$, where x denotes the number of vases manufactured. The vases are sold at a fixed price of \$25 per vase.

For how many vases manufactured by the company is the profit made the maximum?

(A) 10

(B) 13

(C) 18

(D) 21

215. In a laboratory experiment, it was found that the number of bacteria triples every hour. The number of bacteria at the end of six hours was found to be 72,900.

If the number of bacteria initially during the start of the experiment be N, which of the following functions best describes the number of bacteria t hours after the start of the experiment?

(A) $N^3 \times t$

(B) $3^N \times t$

(C) N^{3t}

(D) $3^t \times N$

216. In a laboratory experiment, it was found that the number of bacteria triples every hour. The number of bacteria at the end of six hours was found to be 72,900.

How many hours after the experiment started was the number of bacteria one-third of the final number of bacteria, 72,900?

(A) 2

(B) 3

(C) 4

(D) 5

217. In a laboratory experiment, it was found that the number of bacteria triples every hour. The number of bacteria at the end of six hours was found to be 72,900.

What is the count of bacteria with which the experiment was started?

(A) 100

(B) 500

(C) 1,000

(D) 1,500

218. **SPR:**

The number of vehicles plying on Waterloo Street at any hour of a day is given by the function $v(t) = -t^2 + 20t + 100$, with $0 \leq t \leq 24$; where t denotes the number of whole hours past 12:00 midnight.

After how many hours from midnight is the number of vehicles on the street the maximum?

219. **SPR:**

The number of vehicles plying on Waterloo Street at any hour of a day is given by the function $v(t) = -t^2 + 20t + 100$, with $0 \leq t \leq 24$; where t denotes the number of whole hours past 12:00 midnight.

After how many hours from midnight is the number of vehicles on the street the same as that at midnight?

3.4 Geometry & Trigonometry

3.4.1 Geometry

220.

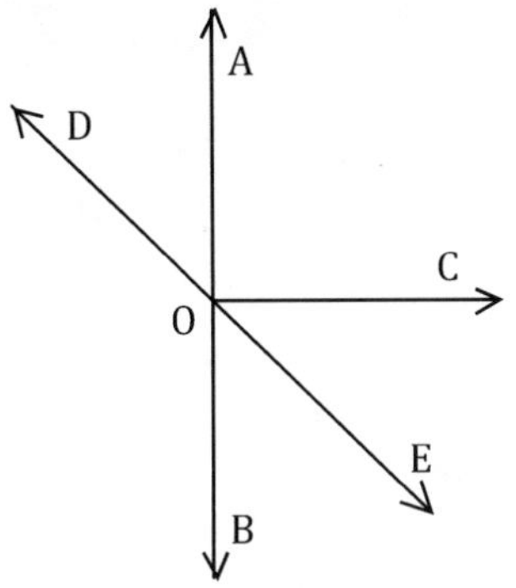

In the figure above, lines AB and DE intersect at O. Ray OC is perpendicular to AB. If angle AOD is twice of angle COE, what is the measure of angle EOA?

(A) 30°

(B) 60°

(C) 120°

(D) 150°

221.

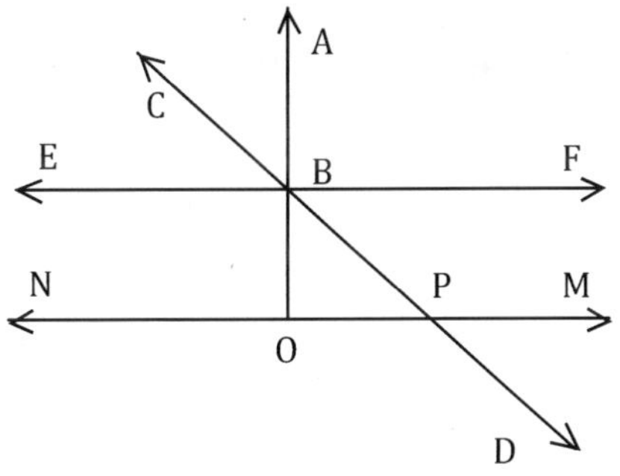

In the figure above, lines MN and CD intersect at P. Ray OA is perpendicular to MN. Lines EF, CD and ray OA intersect at B. If lines MN and EF are parallel and $\angle$MPD is 20° more than $\angle$ABC, what is the measure of $\angle$CBE?

(A) 35°

(B) 45°

(C) 55°

(D) 65°

222.

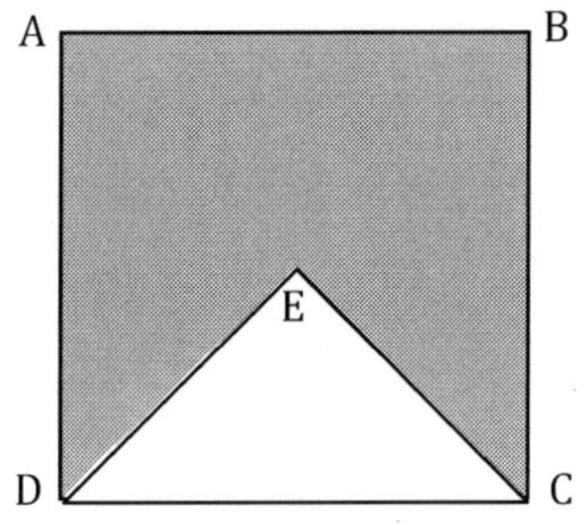

SPR:

ABCD is a square of side 2 units. The height of the triangle DEC inside the square is 1 unit. What is the area (in square units) of the shaded region?

223.

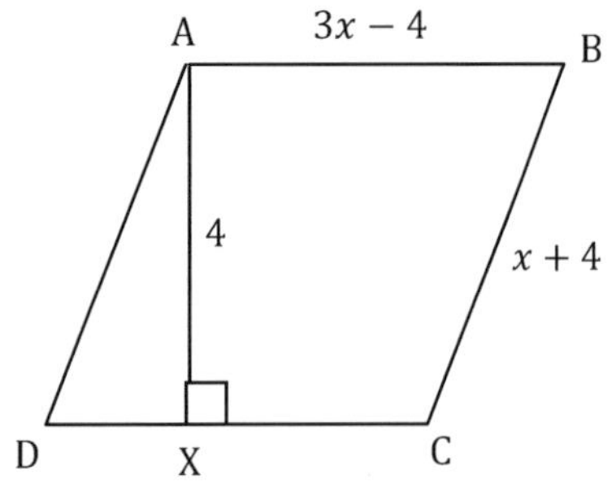

SPR:

In the above parallelogram ABCD of perimeter 24 units, the sides have been mentioned in terms of x. If AX, the perpendicular from A on CD is 4 units, what is the area (in square units) of ABCD?

224.

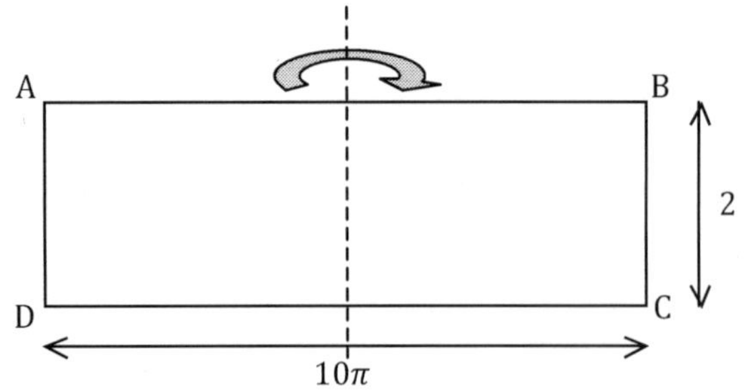

A rectangular sheet of paper ABCD is folded as shown in the figure above to form an open cylinder so that the edges BC and AD overlap. If AB = 10π units and BC = 2 units, what is the volume (in cubic units) enclosed by the above cylinder?

(A) 50

(B) 25π

(C) 50π

(D) 100π

225.

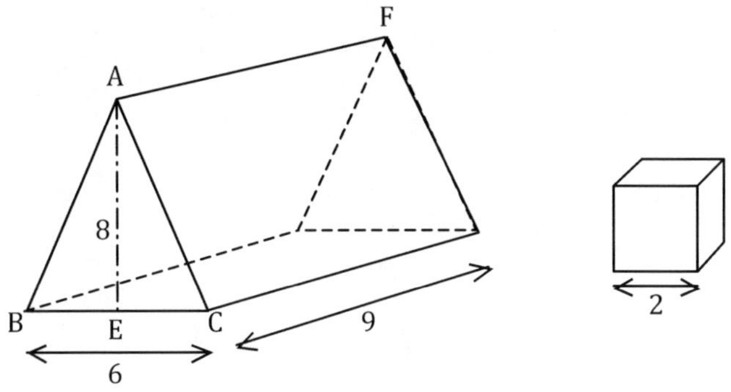

The triangular prism shown above is made of iron. The triangle ABC has base BC of 6 units and height AE as 8 units. The length of the prism, AF is 9 units. It is melted and recast into small identical cubes (shown above) having sides of 2 units each. How many such cubes can be made if there is no loss of metal during recasting?

(A) 8

(B) 27

(C) 36

(D) 54

226.

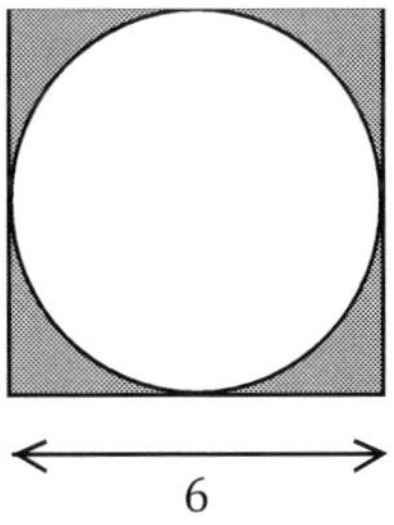

The square above has sides of 6 units. A circle in inscribed in the square touching the four sides. What is the area of the shaded region?

(A) $36(\pi - 1)$

(B) $24 - 9\pi$

(C) $18(4 - \pi)$

(D) $9(4 - \pi)$

227. What is the volume (in cubic units) of the largest sphere that can be placed completely inside a cylinder of radius 2 units and height 3 units?

(A) $\frac{4\pi}{3}$

(B) $\frac{9\pi}{2}$

(C) $\frac{32\pi}{3}$

(D) 36π

228. The total surface area of a cube is $96k$. What is the volume of the cube?

(A) $64k$

(B) $64k\sqrt{k}$

(C) $64k^2$

(D) $384k\sqrt{6k}$

229. The radius of a cylinder is three times the radius of a sphere. If the volume of the sphere is double that of the cylinder, what is the ratio of the height to the radius of the cylinder?

(A) $2:81$

(B) $4:81$

(C) $2:9$

(D) $4:9$

230.

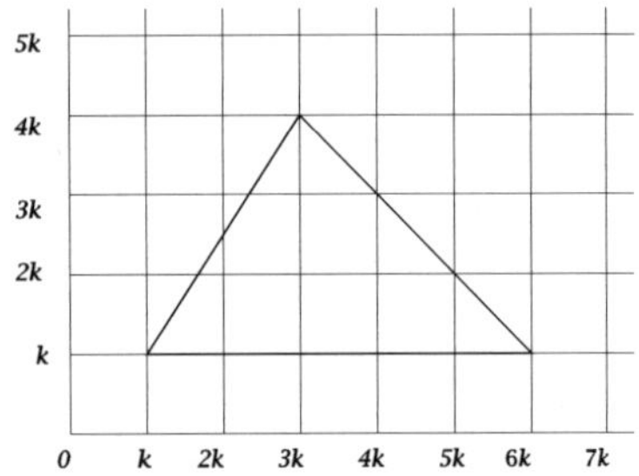

If the area of the triangle above is 35, what is the value of k?

(A) $\sqrt{2}$

(B) $\frac{\sqrt{21}}{3}$

(C) $\sqrt{3}$

(D) $\frac{\sqrt{42}}{3}$

231.

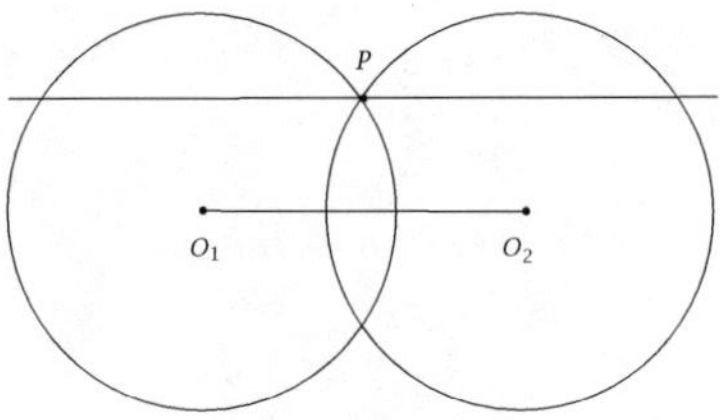

In the figure above, two circles with radii equal to 10 units of length overlap such that their centers, O_1 and O_2, are 16 units apart. What is the distance between the line joining the centers of the circles and the line drawn through point P parallel to the line connecting the centers?

(A) 3

(B) 5

(C) 6

(D) 8

232.

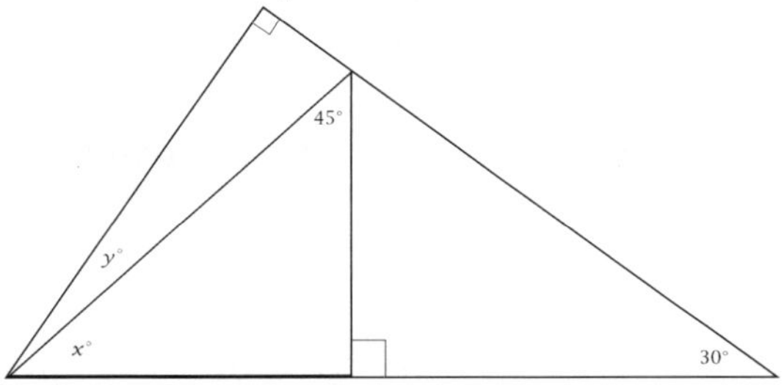

Note: Figure not drawn to scale

In the figure above, what is the value of $(2x - y)$? x and y are denoted as the degrees of the corresponding angles.

(A) 30

(B) 45

(C) 60

(D) 75

233.

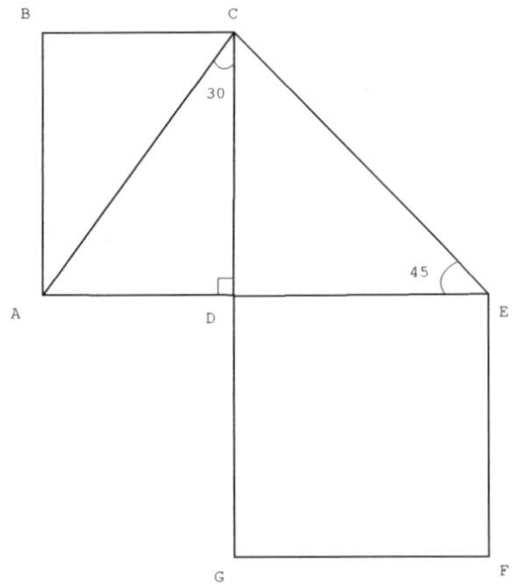

If the area of rectangle $ABCD$ is $4\sqrt{3}$, what is the area of the square $DEFG$?

(A) $2\sqrt{3}$

(B) 4

(C) $4\sqrt{3}$

(D) 12

234. The total distance between points A and C is 150 kilometers. Point B lies between these two points. If the distance between A and B is five times as long as that between B and C, what is the distance between points B and C?

(A) 18

(B) 20

(C) 22

(D) 25

235.

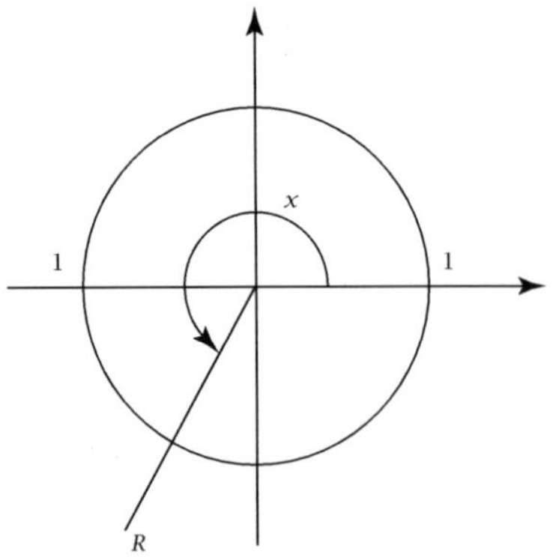

Ray R starts its counter-clockwise revolution around the origin from the positive direction of the X-axis in a coordinate plane. After it travels for one full revolution and an additional angle of $x°$, as shown in the diagram, it stops at the above-illustrated position in the third quadrant. Let D denote the total distance traveled by a point $(1, 0)$ on ray R. Which statement(s) about D are true?

I. $\frac{5\pi}{4} < D < \frac{3\pi}{2}$

II. $\frac{\pi}{2} < D < \pi$

III. $3\pi < D < \frac{7\pi}{2}$

(A) I only

(B) II only

(C) III only

(D) I and III only

236.

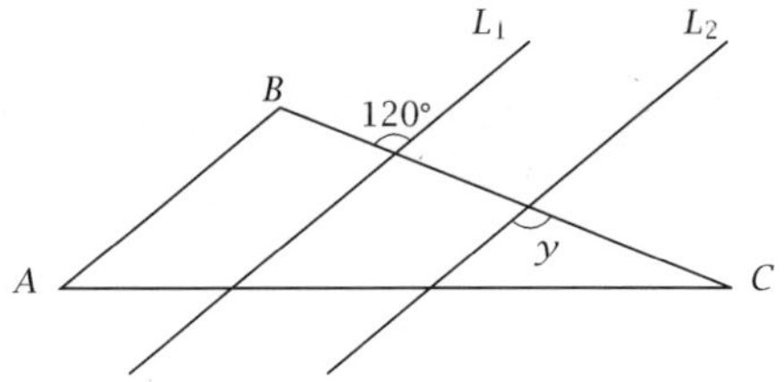

In $\triangle ABC$ above, $AB \| L_1$ & $L_1 \| L_2$. What is the measure of angle y?

(A) 80°

(B) 90°

(C) 100°

(D) 120°

237.

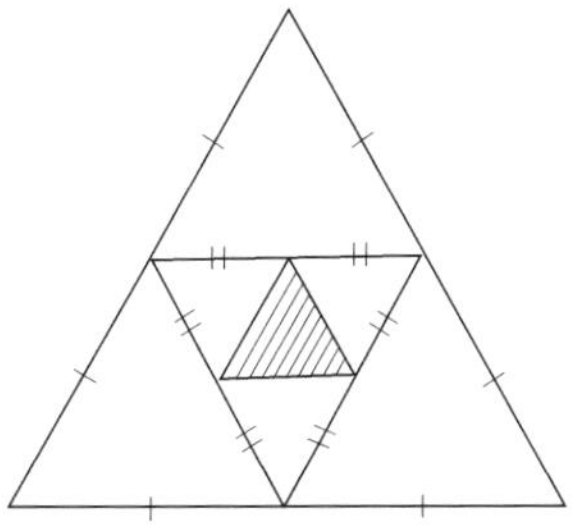

In the figure above each triangle is equilateral, what fraction of the largest triangular region is shaded?

(A) $\frac{1}{16}$

(B) $\frac{1}{9}$

(C) $\frac{3}{16}$

(D) $\frac{1}{4}$

238.

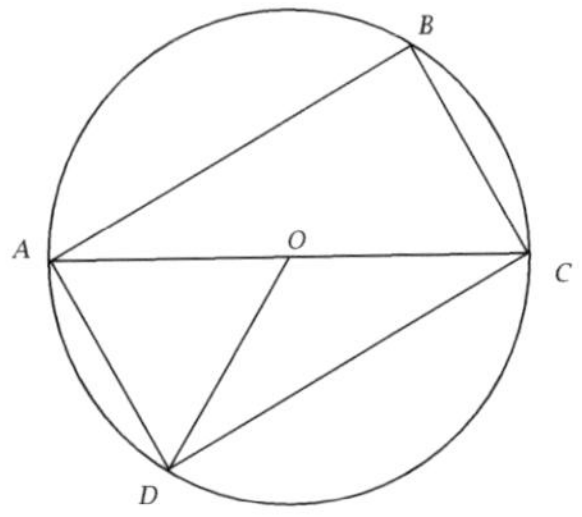

A rectangle is inscribed in the circle in the figure drawn above. The circle has a center at point O and a radius of length 6. The area of $\triangle ABC$ is $18\sqrt{3}$. What is the area of $\triangle ODC$ if points B, D, and O lie on the same line?

(A) 12

(B) $9\sqrt{3}$

(C) $6\sqrt{3}$

(D) 10

239.

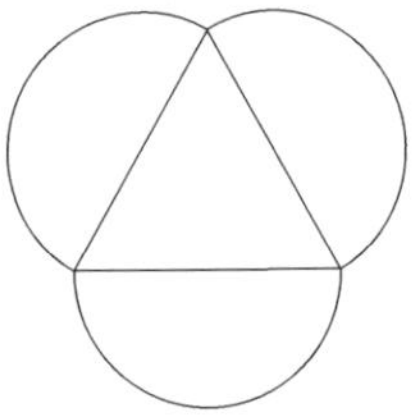

Above is the base of a solid which is 5 centimeters in height. The base consists of an equilateral triangle and three semi-circles on each side of the triangle. Each side of the triangle is 40 centimeters. What is the volume in cubic centimeters of the solid?

(A) $2,400\pi + 400\sqrt{3}$

(B) $3,000\pi + 2,000\sqrt{3}$

(C) $12,000\pi + 2,000\sqrt{3}$

(D) $14,400\pi$

240.

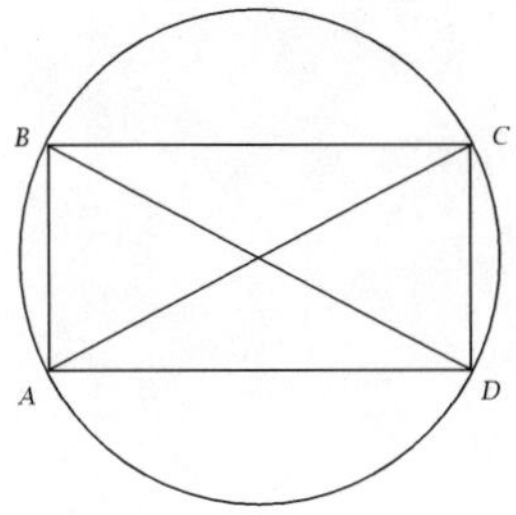

Rectangle $ABCD$ with a perimeter of 60 is inscribed in a circle with a radius $\frac{15}{\sqrt{2}}$ units. What is the area of $ABCD$? (Note: Figure not drawn to scale.)

(A) 150

(B) 225

(C) 450

(D) 750

241. The number of diagonals in a polygon having n sides is given be the formula $\frac{n(n-3)}{2}$. What is the value of n if it is known that the number of diagonals is double the number of sides of the polygon?

(A) 4

(B) 5

(C) 7

(D) 8

242. For a rectangle, the difference between the length and breadth is 4 inches. If the length was increased by 6 inches, keeping the breadth same, the area increased by 120 square inches. What is the breadth of the rectangle?

(A) 10

(B) 20

(C) 24

(D) 30

243. When the side of a square was doubled, the area of the square increased by 300 square inches. What was the initial length (in inches) of a side of the square?

(A) 10

(B) $10\sqrt{2}$

(C) $10\sqrt{3}$

(D) 20

244. A cylinder of length 84 cm is melted down and re-cast into a solid sphere of radius 7 cm. What is the radius of the original cylinder?

(A) $\frac{7}{6}$ cm

(B) $\frac{7}{3}$ cm

(C) $\frac{14}{3}$ cm

(D) 7 cm

245. The sides of a triangle are $2x$, $(x - 1)$, and $(6 - x)$, where x is an integer. What is the perimeter of the triangle?

(A) 5

(B) 7

(C) 9

(D) 10

3.4.2 Co-ordinate Geometry

246. Which of the following represents the area of the region denoted by $2x - y \geq 0$, $x + y \leq 6$ and $y \geq 0$?

(A)

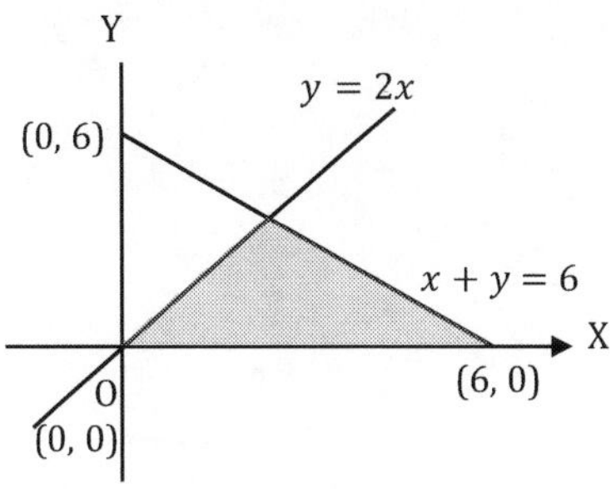

(B)

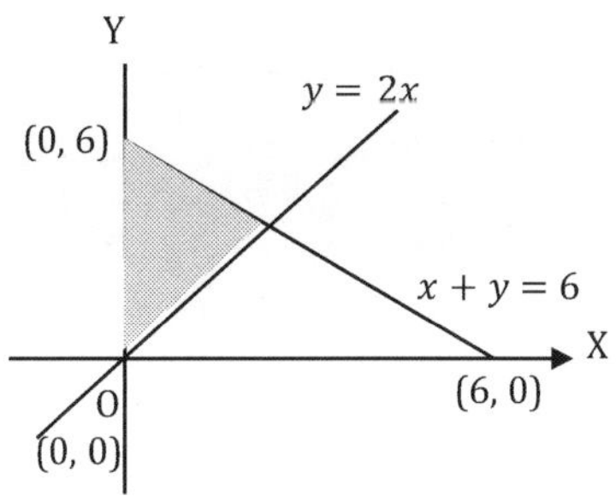

(C)

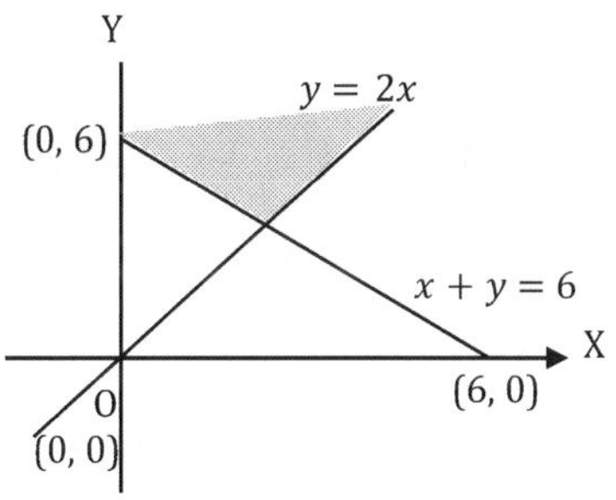

(D)

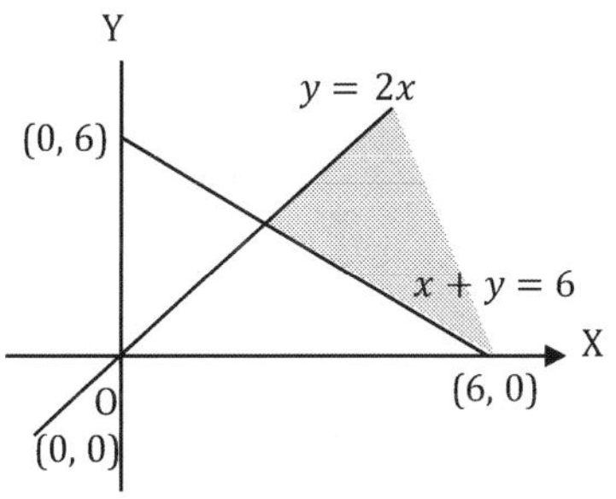

247. SPR:

$x^2 - 4x + y^2 + 6y = 12$

The equation above denotes a circle having center at (p, q) and radius r. What is the value of $(p + q + r)$?

248. What is the equation of the circle inscribed in a square with one of its diagonals having the coordinates (3, 1) and (11, 7)?

(A) $(x+7)^2 + (y-4)^2 = 10$

(B) $(x+7)^2 + (y+4)^2 = 25$

(C) $2(x-7)^2 + 2(y-4)^2 = 50$

(D) $2(x-7)^2 + 2(y-4)^2 = 25$

249. If the lines $\frac{x}{a} + \frac{y}{b} = 1$ and $\frac{x}{p} + \frac{y}{q} = 1$ are perpendicular to one another, which of the following is correct?

(A) $ap - bq = 0$

(B) $ap + bq = 0$

(C) $ap + bq = 1$

(D) $ap - bq = 1$

250. SPR:

Lattice points are those points whose coordinates can take integral values. What is the number of lattice points on the boundary and inside the region bounded by a circle having the equation $x^2 + y^2 = 4$?

251. A and B are two points with the co-ordinates $(-3, 0)$ and $(0, 4)$, respectively. What is the length of the diagonal AC, if AB forms one of the sides of the square ABCD?

(A) 5

(B) 6

(C) $5\sqrt{2}$

(D) 10

252. If the points $(1, 1)$, $(a, 0)$ and $(0, b)$ are collinear, which of the following is equivalent to $(a + b)$?

(A) $\frac{a}{b}$

(B) $\frac{b}{a}$

(C) $(a+1)(b+1)$

(D) ab

253. **SPR:**

If the vertices of a triangle have the co-ordinates $(0,0)$, $(3,2)$ and $(0,5)$, what is the area of the triangle?

254. If the points A $(7,\ 9)$, B $(3,\ -7)$ and C $(-3,\ 3)$ are the vertices of a triangle, what is the measure of $\angle C$?

(A) 30°

(B) 45°

(C) 60°

(D) 90°

255. What is the ratio in which the Y axis intersects the line joining the points $(4,5)$ and $(10,2)$?

(A) $1:3$

(B) $5:2$

(C) $1:1$

(D) $3:2$

3.4.3 Trigonometry

256. In the figure shown below, ABC is a triangle, right angled at B. What is the length of AC?

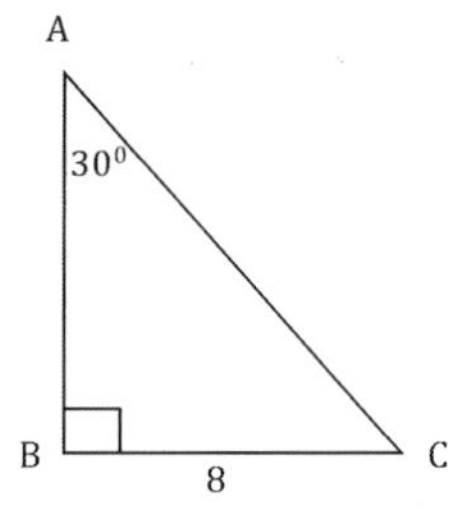

(A) 4

(B) $\frac{16}{\sqrt{3}}$

(C) 16

(D) $16\sqrt{3}$

257. In the figure shown below, ABC is a triangle right angled at B. What is the length of AC?

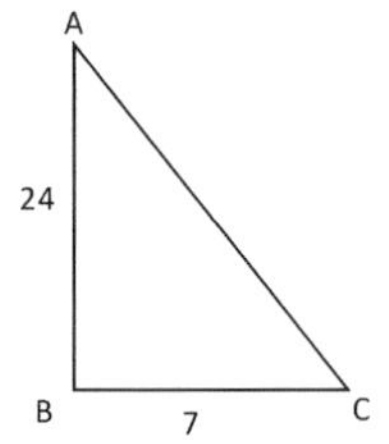

(A) 17

(B) $\sqrt{527}$

(C) 25

(D) 31

258. In a right angled isosceles triangle, the length of the hypotenuse is 8. What is the sum of the lengths of the other two sides?

(A) 4

(B) $4\sqrt{2}$

(C) 8

(D) $8\sqrt{2}$

259. In the figure shown below, ABC is a triangle right angled at B. What is the difference between AC and AB?

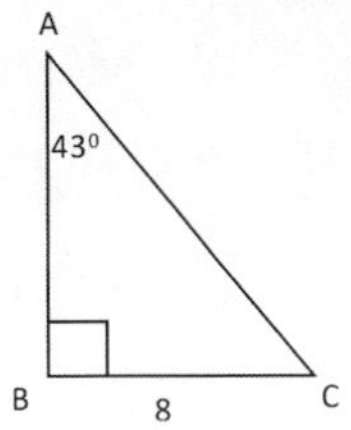

(A) $\frac{8}{\text{Sin } 43^\circ} - \frac{8}{\tan 43^\circ}$

(B) $\frac{8}{\text{Sin } 43^\circ} - \frac{8}{\text{Cos } 43^\circ}$

(C) $\frac{8}{\text{Cos } 43^\circ} - \frac{8}{\tan 43^\circ}$

(D) $\frac{8}{\tan 43^\circ} - \frac{8}{\text{Sin } 43^\circ}$

260. In the figure shown below, ABC is a triangle right angled at B. Through D, a perpendicular is dropped on BC so that BE = EC. If it is known that BC = 8, find the length of DC.

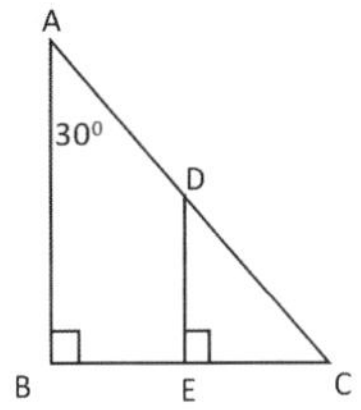

(A) $4\sqrt{2}$

(B) 8

(C) $8\sqrt{2}$

(D) $8\sqrt{3}$

261. In the figure shown below, ABC is a triangle right angled at B. What is the value of angle A?

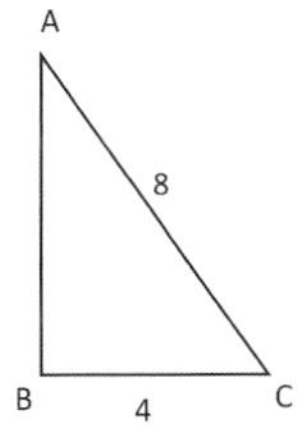

(A) 30°

(B) 45°

(C) 60°

(D) 75°

262. Matt has a garden in the shape of a right angled triangle with one of the acute angles as 30°. If the longest side of the triangle is 8 m long, what is the perimeter of the garden?

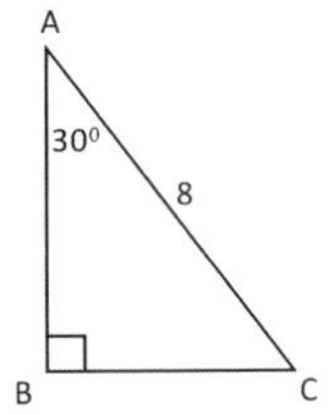

(A) $4\sqrt{3}$

(B) $12 + \frac{4}{\sqrt{3}}$

(C) $8 + 4\sqrt{3}$

(D) $12 + 4\sqrt{3}$

263. Andrew drew a right angled triangle with one angle as 30^0. If the length of the smallest side is 4, what is the area of the triangle?

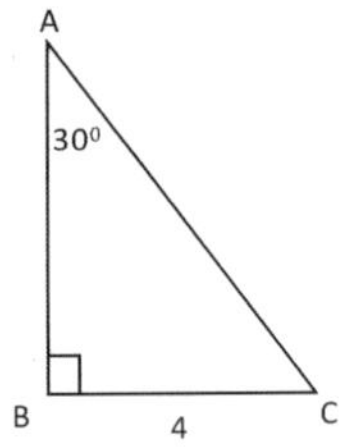

(A) $\frac{8}{\sqrt{3}}$

(B) $8\sqrt{3}$

(C) $12\sqrt{3}$

(D) $16\sqrt{3}$

264. In the figure shown beside, ABC is a triangle, right angled at B. Through B, a line is drawn perpendicular to AC which meets AC in D. What is the length of BD?

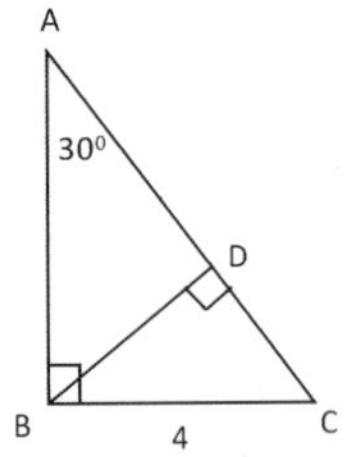

(A) $\frac{2}{\sqrt{3}}$

(B) $\frac{4}{\sqrt{3}}$

(C) $2\sqrt{3}$

(D) $4\sqrt{3}$

265. A company decided to launch a promotional campaign using a hot air balloon. The balloon was tied to a point on the ground using a rope 30 meter long. One day, because of a strong wind, the rope made an angle of 60° with the ground. How high (in meters) was the balloon above the ground level that day?

(A) $\frac{15}{\sqrt{3}}$

(B) 15

(C) $\frac{30}{\sqrt{3}}$

(D) $15\sqrt{3}$

266. What is the value of $(\frac{3\pi}{4}^c + \frac{11\pi}{5}^c - \frac{7\pi}{10}^c)$, when converted to degrees?

(A) 135°

(B) 245°

(C) 405°

(D) 810°

267. Which of the following is the correct value of $(\frac{7\pi}{4}^c + \frac{4\pi}{3}^c - \frac{5\pi}{12}^c - 180^\circ)$?

(A) $\frac{5\pi}{4}^c$

(B) $\frac{5\pi}{3}^c$

(C) 240°

(D) 450°

268. Which of the following statement(s) is/are correct about the angle $\frac{11\pi}{3}$ radians?

I. It is equivalent to 660°.

II. The angle falls in the third quadrant.

III. Cos $\frac{11\pi}{3}$ is positive.

(A) Only I

(B) Only II

(C) Both I and II

(D) Both I and III

269. Which of the following statement(s) is/are correct about the angle $\frac{13\pi}{4}$ radians?

I. It is equivalent to $585°$.

II. The value of tan $\frac{13\pi}{4}$ is -1.

III. The value of Cos $\frac{13\pi}{4}$ is $\frac{1}{\sqrt{2}}$.

(A) Only I

(B) Only II

(C) Both I and II

(D) Both I and III

270. Which of the following statement(s) is/are correct for the angle θ (in radians) if it is known that Cos $\theta = \frac{\sqrt{3}}{2}$?

I. The angle falls in the first quadrant or the second quadrant.

II. The value of tan θ could be negative.

III. The value of Sin θ must be positive.

(A) Only I

(B) Only II

(C) Both I and III

(D) Both II and III

271. An ant started moving in the anti-clockwise direction around the unit circle from the point A as shown below. At a point of time, it stopped at a point after covering a total angle of $\frac{7\pi}{6}^{c}$. What could be the possible co-ordinates of the point on the circle where the ant stopped?

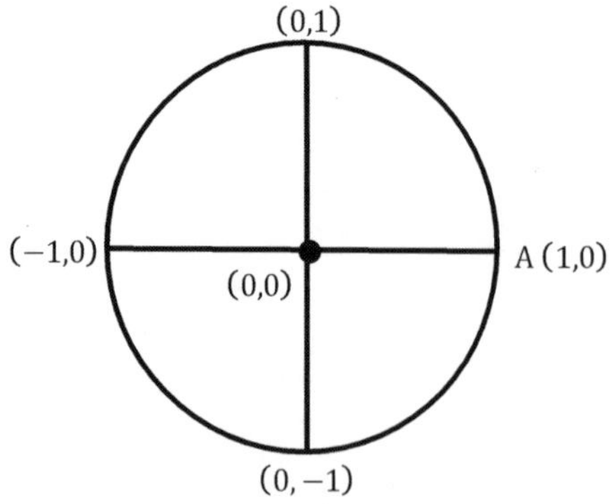

(A) $\left(\frac{1}{\sqrt{2}}, \frac{1}{\sqrt{2}}\right)$

(B) $\left(\frac{\sqrt{3}}{2}, -\frac{1}{2}\right)$

(C) $\left(-\frac{\sqrt{3}}{2}, \frac{1}{2}\right)$

(D) $\left(-\frac{\sqrt{3}}{2}, -\frac{1}{2}\right)$

272. Andrew started tracing out the circumference of the unit circle from the point A as shown below in the clockwise direction. He stopped after tracing out an angle of $\frac{5\pi}{4}^c$. What could be the possible co-ordinates of the point on the circle where he stopped?

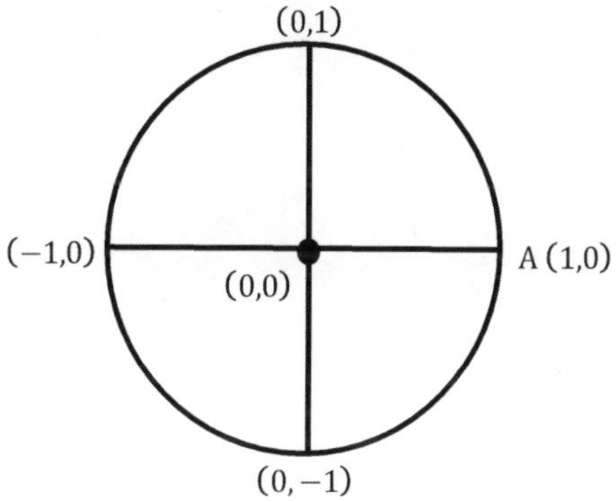

(A) $(\frac{1}{\sqrt{2}}, \frac{1}{\sqrt{2}})$

(B) $(-\frac{1}{\sqrt{2}}, \frac{1}{\sqrt{2}})$

(C) $(\frac{1}{\sqrt{2}}, -\frac{1}{\sqrt{2}})$

(D) $(-\frac{1}{\sqrt{2}}, -\frac{1}{\sqrt{2}})$

273. If the angle θ satisfies the condition that $\frac{4\pi}{5} \leq \theta \leq \frac{5\pi}{6}$, then which of the following statement(s) is/are correct?

I. The value of Cos θ is positive.

II. The value of Cos (2θ) is positive.

III. The value of Sin (2θ) is negative.

(A) Only II

(B) Both I and II

(C) Both II and III

(D) Both I and III

274. If the angle θ satisfies the condition that $\frac{3\pi}{4} < \theta < \frac{4\pi}{5}$, then which of the following statement(s) is/are correct?

I. The value of $\tan \theta$ is negative.

II. The value of $\tan (2\theta)$ is negative.

III. The value of $\tan (3\theta)$ is positive.

(A) Only I

(B) Only III

(C) Both II and III

(D) I, II and III

275. If $\theta = 150°$, then which of the following statement(s) is/are correct?

I. $\text{Sin}\ (2\theta)$ and $\text{Cos}\ (4\theta)$ have the same sign.

II. $\tan\ (2\theta)$ and $\text{Cos}\ (5\theta)$ have opposite signs.

III. $\text{Sin}\ (4\theta)$ and $\tan\ (5\theta)$ have the same sign.

(A) Only I

(B) Only III

(C) Both I and II

(D) None of the statements

Chapter 4

Answer-key

Answer Key

(1) C

(2) A

(3) C

(4) A

(5) D

(6) 7/15 or 0.46 or 0.47

(7) 5

(8) 9/20

(9) 1/2

(10) A

(11) A

(12) 13/3 or 4.33.

(13) 1

(14) B

(15) C

(16) A

(17) 4

(18) 5

(19) C

(20) D

(21) D

(22) D

(23) 22/5 or 4.4

(24) C

(25) C

(26) B

(27) 20

(28) C

(29) A

(30) A

(31) D

(32) A

(33) C

(34) B

(35) 10

(36) C

(37) C

(38) D

(39) C

(40) B

(41) A

(42) $x = 5$ or $-7/3$

(43) $x = 3$ or $-31/3$

(44) D

(45) D

(46) C

(47) D

(48) D

(49) 30

(50) A

(51) D

(52) B

(53) A

(54) A

(55) A

(56) C

(57) $x = 7/6$

(58) 8

(59) C

(60) C

(61) B

(62) 33.3

(63) D

(64) D

(65) C

(66) D

(67) B

(68) B

(69) B

(70) 220

(71) 14.3

(72) B

(73) C

(74) B

(75) C

(76) 2/11 or 0.18

(77) 14.9

(78) 60

(79) B

(80) D

(81) B

(82) D

(83) C

(84) D

(85) B

(86) A

(87) 75

(88) C

(89) 3/34 or 0.09

(90) 3/19 or 0.16

(91) 20

(92) D

(93) 12.5

(94) C

(95) B

(96) D

(97) A

(98) B

(99) C

(100) 250

(101) 34

(102) 28.6

(103) 10.5

(104) 20

(105) 37.5

(106) C

(107) C

(108) 222

(109) 25

(110) A

(111) B

(112) A

(113) C

(114) C

(115) D

(116) A

(117) B

(118) B

(119) D

(120) C

(121) D

(122) D

(123) C

(124) B

(125) B

(126) B

(127) 111

(128) 20

(129) 114

(130) D

(131) D

(132) C

(133) B

(134) B

(135) B

(136) B

(137) 1/12

(138) 1/8

(139) 3/8

(140) C

(141) C

(142) B

(143) D

(144) C

(145) C

(146) D

(147) C

(148) 151

(149) 18

(150) A

(151) D

(152) D

(153) B

(154) 4.97

(155) C

(156) D

(157) D

(158) D

(159) C

(160) A

(161) C

(162) D

(163) D

(164) A

(165) D

(166) C

(167) C

(168) B

(169) B

(170) A

(171) 7

(172) B

(173) 375

(174) D

(175) C

(176) B

(177) B

(178) B

(179) B

(180) C

(181) C

(182) D

(183) 2

(184) C

(185) D

(186) D

(187) D

(188) D

(189) C

(190) B

(191) B

(192) D

(193) D

(194) C

(195) D

(196) A

(197) D

(198) C

(199) D

(200) D

(201) 7

(202) 36

(203) A

(204) A

(205) 3

(206) B

(207) D

(208) 17

(209) B

(210) D

(211) B

(212) C

(213) B

(214) D

(215) D

(216) D

(217) C

(218) 10

(219) 20

(220) C

(221) C

(222) 3

(223) 20

(224) C

(225) B

(226) D

(227) B

(228) B

(229) A

(230) D

(231) C

(232) D

(233) D

(234) D

(235) C

(236) D

(237) A

(238) B

(239) B

(240) B

(241) C

(242) B

(243) 4

(244) B

(245) C

(246) A

(247) 4

(248) D

(249) B

(250) 13

(251) C

(252) D

(253) 7.5

(254) D

(255) B

(256) C

(257) C

(258) D

(259) A

(260) B

(261) A

(262) D

(263) B

(264) C

(265) D

(266) C

(267) B

(268) D

(269) A

(270) B

(271) D

(272) B

(273) C

(274) D

(275) C

Chapter 5

Solution

5.1 Algebra

5.1.1 Algebraic Expressions

1. We note that the denominators of all the fractions are multiples of 2 and 3 and have an LCM as 6.

We can write the value of m as:

$m = (\frac{5}{3} + \frac{5}{6} + \frac{1}{2}) = \frac{5\times2}{3\times2} + \frac{5}{6} + \frac{1\times3}{2\times3} = \frac{10}{6} + \frac{5}{6} + \frac{3}{6} = \frac{18}{6} = 3.$

Thus, $(m - 3)^3 = (3 - 3)^3 = 0$.

The correct answer is Option C.

2. We know that $150 < Q < 160$

Or, $150 < 17m < 160$

Dividing throughout by 17: $\frac{150}{17} < m < \frac{160}{17}$ => $8.82 < m < 9.41$

Since m is an integer, the only value qualifies for $m = 9$.

The correct answer is Option A.

3. We know that $15 < m < 23$ => $16 \leq m \leq 22$ (since m is an integer).

Also, $Q = 17m$

So, the minimum value of $Q = 16 \times 17 = 272$ and the maximum value of $Q = 22 \times 17 = 374$.

=> $272 \leq Q \leq 374$.

Hence, options A and D are eliminated. The correct option is between B and C.

The correct option will be that number which is divisible by 17.

Option C (= 306) is divisible by 17, while Option B (= 290) is not.

Note: We do not need to check both B and C. We check any one option. If that option is divisible by 17, it must be the answer, since there cannot be more than one correct option. If, however, that option turns out not to be divisible by 17, then the other option must be correct.

The correct answer is Option C.

4. We have

$$N = \frac{S}{\frac{p}{q} - 1}$$

$=> N = \dfrac{S}{\frac{p-q}{q}}$

$=> N = \dfrac{Sq}{p-q}$

$=> Np - Nq = Sq => q(N+S) = Np$

$=> q = \dfrac{Np}{N+S}$

The correct answer is Option A.

5. $\dfrac{3}{\frac{3}{4}-\frac{5}{12}} = \dfrac{3}{\frac{3\times 3}{4\times 3}-\frac{5}{12}} = \dfrac{3}{\frac{9}{12}-\frac{5}{12}} = \dfrac{3}{\frac{9-5}{12}} = \dfrac{3}{\frac{4}{12}} = 3\times\dfrac{12}{4} = 3\times 3 = 9.$

The correct answer is Option D.

6. We know that $p = \dfrac{1+\frac{x}{y}}{1+\frac{xy}{z^2}}$.

We see that x and y are expressed in terms of z. So, if we substitute the value of x and y in the ratio $\frac{x}{y}$, we will get a numerical value (since z will get eliminated). Similarly, for the ratio $\frac{xy}{z^2}$, if we substitute the values of x and y, then z^2 will be eliminated and we will get a numerical value.

Let us calculate the value of $\frac{x}{y}$ and plug in the above expression.

Now, $\frac{x}{y} = \frac{\frac{z}{2}}{3z} = \frac{z}{2}\times\frac{1}{3z} = \frac{1}{6}$, by replacing the given values of x and y.

Calculating the value of $\frac{xy}{z^2}$, a term of the denominator of the expression for p...

$\dfrac{xy}{z^2} = \dfrac{[(\frac{z}{2})\times 3z]}{z^2} = \dfrac{(\frac{3z^2}{2})}{z^2} = \dfrac{3}{2}.$

Thus, we have $p = \dfrac{(1+\frac{1}{6})}{(1+\frac{3}{2})} = \dfrac{7}{6}\times\dfrac{2}{5} = \dfrac{7}{15}.$

The correct answer is 7/15 or 0.46 or 0.47 (rounded off).

7. $a = 216 = 6^3$; representing in exponential form

$=> a^{\frac{1}{3}} = (6^3)^{\frac{1}{3}} = 6^{3\times\frac{1}{3}} = 6.$

Thus, $a^{\frac{1}{3}} - 1 = 6 - 1 = 5.$

The correct answer is 5.

8. $\left(\frac{1}{x-9}\right)^n + \left(\frac{1}{x}\right)^{1-n} = \left(\frac{1}{25-9}\right)^{\frac{1}{2}} + \left(\frac{1}{25}\right)^{1-\frac{1}{2}} = \left(\frac{1}{16}\right)^{\frac{1}{2}} + \left(\frac{1}{25}\right)^{\frac{1}{2}} = \frac{1}{4} + \frac{1}{5} = \frac{9}{20}.$

Thus, the value of the given expression is $\frac{9}{20}$.

The correct answer is 9/20 or 0.45.

9. we have $\frac{a^2+b^2}{(a+b)^2+(a-b)^2} = \frac{a^2+b^2}{(a^2+2ab+b^2)+(a^2-2ab+b^2)} = \frac{a^2+b^2}{2(a^2+b^2)} = \frac{1}{2}$. [By expanding $(a+b)^2$ & $(a-b)^2$]

Thus, the value of the expression is $\frac{1}{2}$ irrespective of the values of a and b.

Hence, the answer is $\frac{1}{2}$.

A traditional approach to solve this question would have been time consuming.

Let us see how.

By replacing the values of a & b in the expression, we get,

$\frac{a^2+b^2}{(a+b)^2+(a-b)^2} = \frac{2.3^2+1.7^2}{(2.3+1.7)^2+(2.3-1.7)^2} = \frac{5.29+2.89}{4^2+0.6^2} = \frac{8.18}{16+0.36} = \frac{1}{2}$.

The correct answer is 1/2.

10. $\frac{x^2-12x+32}{x-4} = \frac{x^2-4x-8x+32}{x-4} = \frac{x(x-4)-8(x-4)}{x-4} = \frac{(x-4)(x-8)}{x-4} = (x-8)$.

Hence, the value of the expression $x - 8 = 6.5 - 8 = -1.5$.

[Note: One could also have simply plugged in the value of $x = 6.5$ in the expression and calculated the value. However, the calculation would have been tedious.]

The correct answer is Option A.

11. $\frac{x^2-12x+32}{x+4} = \frac{x^2-4x-8x+32}{x+4} = \frac{x(x-4)-8(x-4)}{x+4} = \frac{(x-4)(x-8)}{x+4}$

$= \frac{(-6.5-4)(-6.5-8)}{(-6.5+4)} = \frac{(-10.5)\times(-14.5)}{-2.5} = -60.9$.

The correct answer is Option A.

12. We need to calculate the value of 1#1#5 performing the operations from left to right.

So, we have $(1\#1)\#5 = (\frac{1^2+1^2}{2})\#5 = \frac{2}{2}\#5 = 1\#5 = \frac{1^2+5^2}{1+5} = \frac{26}{6} = \frac{13}{3} = 4.33$.

The correct answer is 13/3 or 4.33.

13. The mixed fraction $Z\frac{Z}{3}$ can be manipulated as a proper fraction, and is equal to $= \frac{3Z+Z}{3} = \frac{4Z}{3}$.

Thus, we have $\frac{4Z}{3} = \frac{8}{3} => Z = \frac{8}{3} \times \frac{3}{4} = 2$.

Hence, the value of $\frac{1}{2Z-3} = \frac{1}{2\times2-3} = \frac{1}{4-3} = 1$.

The correct answer is 1.

5.1.2 Numbers

14. We know the following results for summation:

- Even + Even = Even
- Odd + Odd = Even
- Even + Odd = Odd

In general, the following are valid:

The sum of any number of even terms is even

The sum of even number of odd terms is even

The sum of odd number of odd terms is odd

For product, we have the following results:

- Even × Even = Even
- Odd × Odd = Odd
- Even × Odd = Even

In general, an even number multiplied with any integer (even or odd) always results in an even number.

Statement I: In case of 5 consecutive integers, two cases are possible:

(1) There are 3 odd numbers and 2 even numbers:
Here, sum of the 3 odd numbers is odd and the sum of the even numbers is even. So, the total is the sum of one odd number and one even number, which is odd (example: 3, 4, 5, 6, 7).

(2) There are 2 odd numbers and 3 even numbers:
Here, sum of the 2 odd numbers is even and the sum of the even numbers is even. So, the total is the sum of two even numbers, which is even (example: 2, 3, 4, 5, 6).

So, Statement I is false (since we need the result to be definitely odd).

Statement II: In case of 14 consecutive integers, there will always be 7 odd numbers and 7 even numbers. Here, sum of the 7 odd numbers is odd and the sum of the even numbers is even. So, the total is the sum of one odd number and one even number which is odd (example: 3, 4, 5, 6, 7 . . . 15, 16).

So, Statement II is always true.

Statement III: When we consider 11 consecutive integers, there would be either 5 odd and 6 even numbers or 6 odd and 5 even numbers. In either case, there would be at least one even number. Hence, the product will always be even.

So, Statement III is false.

The correct answer is Option B.

15. We need to understand the following points:

(1) When we select any 2 consecutive integers, we always have one number even and one number odd.

To validate this, take any 2 consecutive integers: (2, 3), (11, 12), (17, 18), etc.

(2) When we select any 3 consecutive integers, we always have a number–a multiple of 3.

To validate this, take any 3 consecutive integers: (1, 2, 3), (11, 12, 13), (22, 23, 24), etc.

Thus, we can conclude that the product of any three consecutive integers must be a multiple of 3. Also, since three consecutive integers include the case of two consecutive integers as well, the product is also a multiple of 2. Hence, the product is a multiple of 6.

Thus, both statements I and III are correct.

Statement II may or may not be correct.

For example, the product of three numbers in the set (11, 12, 13) is a multiple of 4; however, the product of the numbers in the set (17, 18, 19) is not a multiple of 4.

[Note: In general, we can say: The product of N consecutive positive integers is always divisible by $N!$]

The correct answer is Option C.

16. $11,000 \times a = aa000$ (for example, $11,000 \times 2 = 22,000$, etc.)

Similarly, $10 \times b = b0$

Hence, if we add the numbers, we get:

a	a	0	0	0
	+		b	0
			+	c
a	a	0	b	c

Thus, we get a 5-digit number: $a\ a\ 0\ b\ c$.

The hundred's digit is 0.

The correct answer is Option A.

17. Let us assume that one of the numbers is N and has the digits a and b in its tens and units places, respectively.

Thus, the number N can be expressed as $10a + b$.

Thus, the reversed number is $10b + a$.

Hence, we have $(10a + b) - (10b + a) = 54 => 9(a - b) = 54$

$=> a - b = 6$

Hence, possible values of a and b can be enumerated as:

a	b	$N = 10a + b$
6	0 (minimum value)	60
7	1	71
8	2	82
9 (maximum value)	3	93

Thus, there are 4 such numbers.

The correct answer is 4.

18. We have $\frac{1\#}{3\#} + \frac{4}{7} = 1 => \frac{1\#}{3\#} = 1 - \frac{4}{7} = \frac{7-4}{7} = \frac{3}{7}$.

Thus, we have $\frac{1\#}{3\#} = \frac{3}{7}$.

Now, we need to find two numbers 1# and 3#, having the same unit digits and are in the ratio 3 : 7.

Thus, we need to find a multiple of 3 and a multiple of 7, each having the same unit digit and starts with '1' and '3', respectively. After an observation, we find that the common factor to the multiples must be '5' (since $3 \times 5 = 15$ and $7 \times 5 = 35$).

Hence, the value of # is 5.

The correct answer is 5.

19. The sum of the numbers written down by Bob is:
$P = 1 + 2 + 3 \cdots + 100$.

The sum of the numbers written down by Alfred is:
$Q = 2 + 4 + 6 \cdots + 200 = 2(1 + 2 + 3 \cdots + 100)$.

Thus, we can clearly see that $Q = 2P => P = \frac{Q}{2}$.

The correct answer is option C.

20. Here, the relations between four integers, $a, b, c,$ & d are given. We do not know the actual values of any of the integers. However, from the relations, we can find out the ratio of the integers.

Since $2a = 5b = 3c = 6d$, let us assume that all of these are equal to a multiple of the LCM of 2, 5, 3 and 6. The LCM of 2, 5, 3 and 6 is 30 (Since this is a ratio questions, the choice of the initial number we assume will not affect the final answer. We need to choose a convenient value, making our calculations easy).

So, let us assume that $2a = 5b = 3c = 6d = 30k$, where k is some constant whose value is unknown.

Thus, we have

$a = \frac{30k}{2} = 15k$; $b = \frac{30k}{5} = 6k$; $c = \frac{30k}{3} = 10k$; $d = \frac{30k}{6} = 5k$

Now, we need to find which of the options does not represent an integer.

For this, we substitute the values of $a, b, c,$ & d in the options one-by-one.

Starting with Option A: $\frac{ad}{3} = \frac{15k \times 5k}{3} = 25k^2$.

Now, if the value of k be assumed as an integer, then Option A would result in an integer value.

Option B: $\frac{ac}{d} = \frac{15k \times 10k}{5k} = 30k$.

Now, if the value of k be assumed as an integer, then Option B would result in an integer value.

Option C: $\frac{ac}{bd} = \frac{15k \times 10k}{6k \times 5k} = 5$

Thus, Option C obviously results in an integer value.

Since one of the options must not result in an integer value, the answer must be Option D. We do not need to verify Option D as well.

However, for your curiosity, let us check Option D: $\frac{ab}{cd} = \frac{15k \times 6k}{10k \times 5k} = \frac{9}{5}$

Thus, Option D obviously does not result in an integer value.

The correct answer is option D.

21. Here, the relations between four integers, $a, b, c,$ & d are given. We do not know the actual values of any of the integers.

We need to find which of the options is incorrect. For this, we need to check the options one-by-one.

Starting with Option A: $ab = cd \Rightarrow \frac{a}{c} = \frac{d}{b}$.

Multiplying d to both sides of the equation: $\frac{ad}{c} = \frac{d^2}{b}$

Hence, Option A is true.

Option B: $ab = cd \Rightarrow \frac{a}{c} = \frac{d}{b}$.

Squaring both sides: $\frac{a^2}{c^2} = \frac{d^2}{b^2}$

Adding '1' to both sides: $1 + \frac{a^2}{c^2} = 1 + \frac{d^2}{b^2} \Rightarrow \frac{a^2+c^2}{c^2} = \frac{d^2+b^2}{b^2}$

Hence, Option B is true.

Option C: $ab = cd \ldots$(i)

Adding '1' to both sides: $1 + ab = 1 + cd \ldots$(ii)

Dividing (ii) by (i): $\frac{1+ab}{ab} = \frac{1+cd}{cd} => \frac{b(\frac{1}{b}+a)}{ab} = \frac{c(\frac{1}{c}+d)}{cd} => \frac{\frac{1}{b}+a}{a} = \frac{\frac{1}{c}+d}{d}$

Hence, Option C is true.

Hence, we can conclude that Option D must be incorrect since there must be one option which is not true. We do not need to verify Option D.

However, for your curiosity, let us verify Option D: $ab = cd => \frac{ab}{c} = d$

Multiplying b to both sides:

$=> \frac{ab^2}{c} = bd$

But , Option D says: $\frac{ab^2}{c} = d$

Hence, option D is not true.

The correct answer is Option C.

Alternate approach:

Let us assume some values of $a, b, c,$ & d which satisfy the given relation $ab = cd$.

A set of possible values could be $a = 2, b = 12, c = 3,$ & $d = 8$ (we should try not to use easy values such as $a = b = c = d = 1$ as these tend to satisfy all the options given in such questions). Now, we substitute these values in each option and one-by-one check which of the options is not satisfied.

Option A: $\frac{ad}{c} = \frac{2\times8}{3} = \frac{16}{3}$ and $\frac{d^2}{b} = \frac{8^2}{12} = \frac{64}{12} = \frac{16}{3}$

Thus, both the Left hand side (LHS) and the Right hand side (RHS) give the same value. Hence, Option A is true.

Option B: $\frac{a^2+c^2}{c^2} = \frac{2^2+3^2}{3^2} = \frac{13}{9}$ and $\frac{d^2+b^2}{b^2} = \frac{8^2+12^2}{12^2} = \frac{208}{144} = \frac{13}{9}$

Thus, both the Left hand side (LHS) and the Right hand side (RHS) give the same value. Hence, Option B is true.

Option C: $\frac{\frac{1}{b}+a}{a} = \frac{\frac{1}{12}+2}{2} = \frac{25}{24}$ and $\frac{\frac{1}{c}+d}{d} = \frac{\frac{1}{3}+8}{8} = \frac{25}{24}$

Thus, both the Left hand side (LHS) and the Right hand side (RHS) give the same value. Hence, Option C is true.

Option D: $\frac{ab^2}{c} = \frac{2\times 12^2}{3} = \frac{288}{3} = 96 \neq d$

Thus, both the Left hand side (LHS) and the Right hand side (RHS) do not give the same value. Hence, Option D is not true.

22. Let the initial numbers of green and red apples be $7k$ and $11k$, respectively, and the final numbers of green and green apples are $7k - 10$ and $11k - 10$ respectively.

Since the final ratio is 9 : 17, we have

$$\frac{7k - 10}{11k - 10} = \frac{9}{17}$$

=> $119k - 170 = 99k - 90$

=> $20k = 80$

=> $k = 4$

Thus, the initial number of green apples = $7k = 7 \times 4 = 28$.

The correct answer is Option D.

23. This question is based on the identity: $a^2 - b^2 = (a + b)(a - b)$.

Since $a@b = a^2 - b^2$, we have $x@4 = x^2 - 4^2 = (x + 4)(x - 4)$.

Thus, we have

$(x + 4)(x - 4) = \frac{2x+8}{5}$ => $(x + 4)(x - 4) = \frac{2}{5} \times (x + 4)$

=> $(x + 4)(x - 4) - \frac{2}{5} \times (x + 4) = 0$

Taking $(x + 4)$ common: $(x + 4)\{(x - 4) - \frac{2}{5}\} = 0$

=> $(x + 4)(x - 4 - \frac{2}{5}) = 0$ => $(x + 4)(x - \frac{22}{5}) = 0$.

Since $x \neq -4$, we can say that $x = \frac{22}{5} = 4.4$.

The correct answer is 22/5 or 4.4.

24. Three-digit numbers, yielding remainder '1,' when divided by 100 are $101, 201, 301, 401, 501, 601, 701, 801$ & 901: nine numbers.

Among these nine numbers, we have to select those numbers, which when divided by 60 yield remainder 1; these numbers are $301, 601$, and 901: three numbers.

The correct answer is Option C.

25. Given that $uv = -7$, if product of two integers is negative, one of the integer would be negative and the other would be positive, so u and v are of opposite signs.

If we factorize '7', we get '1×7'; thus u and v can take a numerical value of 1 or 7.

Thus, we have $u = 7$ and $v = -1$ OR $u = -7$ and $v = 1$ OR $u = 1$ and $v = -7$ OR $u = -1$ and $v = 7$.

So, their sum is either 6 or -6, and $6^2 = 36$, or $(-6)^2 = 36$.

The correct answer is Option C.

5.1.3 Factors, Multiples, LCM, & HCF

26. The key to this question is to decompose 48 into its prime factors and use these prime factors to construct the factors of 48 that are greater than $\sqrt{48}$.

Since $7^2 = 49$, $\sqrt{48} < 7$.

We understand that a factor is necessarily an integer, and we are looking for factors that are greater than 7. Prime number decomposition of 48 is $2^4 \times 3$.

To determine a complete set of factors: $2, 2, 2, 2, 3$, we need to work systematically:

- $2 \times 2 = 4$
- $2 \times 3 = 6$
- $2 \times 2 \times 2 = 8$
- $2 \times 2 \times 3 = 12$
- $2 \times 2 \times 2 \times 2 = 16$
- $2 \times 2 \times 2 \times 3 = 24$
- $2 \times 2 \times 2 \times 2 \times 3 = 48$

Thus, the factors of 48 which are more than $\sqrt{48}$ are: $8, 12, 16, 24$, and 48: five.

The correct answer is Option B.

27. The minimum number of squared identical tiles will be used when we use square tiles of the largest possible sides. The key word is 'identical,' i.e., all the tiles must be of the same dimension.

A maximum side of a square tile will be the HCF (Highest Common Factor) of the lengths of the sides of the rectangular floor i.e. the HCF of 48 and 60.

- $48 = 2^4 \times 3$
- $60 = 2^2 \times 3 \times 5$

Hence, the HCF of 48 and 60 = $2^2 \times 3 = 12$ feet.

So, square tiles of size 12 feet by 12 feet should be used.

Thus, number of square tiles required = $\frac{\text{Total area}}{\text{Area of each square}} = \frac{48 \times 60}{12 \times 12} = 20$.

Thus, minimum 20 square tiles need to be used.

The correct answer is 20.

28. Amy writes down every fourth number starting from 1 till 50, hence she writes down the numbers: 1, 5, 9, 13 and so on.

Mary writes down every third number starting from 1 till 50, hence she writes down the numbers: 1, 4, 7, 10, 13 and so on.

Thus, we see that the first number common to both of them after 1 is 13.

We can get the starting number (i.e. 13) in another way:

Since both Amy and Mary start with the same number (1), with Amy counting every fourth number and Mary counting every third number, they would reach a common number after a gap equal to the LCM of 4 and 3 i.e. 12.

Thus, the next number that both reach is: $1 + 12 = 13$.

Thus, in general, they would keep reaching the same numbers always at gaps of 12.

If one observes carefully, the number after 13 which they both reach is: $13 + 12 = 25$.

Thus, the numbers common to both are: 1, 13, 25, 37 and 49.

Hence, they have 5 numbers in common.

Alternately, we can also calculate the # of common numbers in the following way:

We know that the starting number is 1 and the gap between any two consecutive common numbers is 12. Thus, from 1 to 50, i.e. a total gap of $50 - 1 = 49$, we try to find out how many gaps of 12 can be accommodated.

This can be calculated by dividing 49 by 12 and finding the quotient, which, in this case is 4. Thus, there are 4 such gaps possible.

Hence, the # of common numbers will be one more than the number of gaps $= 1 + 4 = 5$.

Hence, the total # of numbers common = 5.

The correct answer is Option C.

29. We need to find the distinct number of prime factors for each of the given numbers when it is multiplied by 5. Then, we need to select the number which has the different number of distinct prime factors. Here, we need to work with each option.

Starting with Option A: $15 = 5^1 \times 3^1$

Thus, $15 \times 5 = 5^2 \times 3^1$;(Multiplying it with '5').

Hence, the number of distinct prime factors are **two**, viz. 3 and 5.

Option B: $28 = 2^2 \times 7^1$

Thus, $28 \times 5 = 2^2 \times 7^1 \times 5^1$

Hence, the number of distinct prime factors are **three**, viz. 2, 5 and 7.

So, we have already got two numbers, 15 & 28, having different number of distinct prime factors. According to the question, only one number will have different number of distinct prime factors. Hence, in the next option, the number will have either **two** or **three**distinct prime factors. Whichever of two or three gets repeated, the other one will be the exception and hence, our answer. So, we would not have to check the fourth number.

Option C: $30 = 2^1 \times 3^1 \times 5^1$.

Thus, $30 \times 5 = 2^1 \times 3^1 \times 5^2$.

Hence, the number of distinct prime factors are **three**, viz. 2, 3 and 5.

Hence, we know that Option A must be the exception since it has **two** distinct prime factors, while options A and C have **three** each. It implies that Option D: 22 must have **three**distinct prime factors as there would be only one option as an exception!

For your curiosity, let us work out Option D: $22 = 2^1 \times 11^1$

Thus, $22 \times 5 = 2^1 \times 11^1 \times 5^1$

Hence, the number of distinct prime are **three**, viz. 2, 11 and 5.

Thus, Option D is not the exception.

The correct answer is Option A.

30. We need to check each number for the highest exponent of any of its prime factors. In order to do that, we need to break each number into its prime form.

Here, we need to work with each option.

Option A:

$48 = 2 \times 2 \times 2 \times 2 \times 3 = 2^4 \times 3^1$.

Thus, the highest exponent for its prime factors is 4 since we have 2^4 as a factor.

Option B:

$72 = 2 \times 2 \times 2 \times 3 \times 3 = 2^3 \times 3^2$.

Thus, the highest exponent for its prime factors is 3 since we have 2^3 as a factor.

Option C:

$250 = 2 \times 5 \times 5 \times 5 = 2^1 \times 5^3$.

Thus, the highest exponent for its prime factors is 3 since we have 5^3 as a factor.

Option D:

$484 = 2 \times 2 \times 11 \times 11 = 2^2 \times 11^2$.

Thus, the highest exponent for its prime factors is 2 since we have 2^2 as well as 11^2 as factors.

Hence, among the given numbers, the one with the highest exponent for any of its prime factors is 48 and the corresponding exponent is 4 (exponent for 2).

The correct answer is Option A.

31. The left hand side of the equation:

$75^n \times 15 = (5^2 \times 3)^n \times 3^1 \times 5^1 = 5^{2n+1} \times 3^{n+1}$.

Comparing this to the right hand side of the equation and equating the exponents, we get:

From the exponents of 3: $n + 1 = 6 => n = 5$.

From the exponents of 5: $2n + 1 = m => m = 2 \times 5 + 1 = 11$.

Thus, the value of $(m + n) = 11 + 5 = 16$.

The correct answer is Option D.

32. Breaking 1,500 into its prime factors:

$1,500 = 2 \times 2 \times 3 \times 5 \times 5 \times 5 = 2^2 \times 3^1 \times 5^3$

The right hand side of the equation:

$6^m \times 50^n \times 5 = (2 \times 3)^m \times (5^2 \times 2)^n \times 5 = 2^m \times 3^m \times 5^{2n} \times 2^n \times 5 = 2^{m+n} \times 3^m \times 5^{2n+1}$

Comparing 1,500 to the right hand side of the equation and equating the exponents, we get:

From the exponents of 2: $m + n = 2$.

From the exponents of 3: $m = 1$.

Thus, $n = 2 - m = 2 - 1 = 1$.

We can verify that the exponent of 5 would match for $n = 1$: $5^{2n+1} = 5^{2\times1+1} = 5^3$.

Thus, the value of $(m - n) = 1 - 1 = 0$.

The correct answer is Option A.

5.1.4 Linear equations

33. We know that $2x = 5y \Rightarrow x = \frac{5y}{2}$.

Hence, substituting the value of x in the equation $x + y = 21$, we get:

$\frac{5y}{2} + y = 21 \Rightarrow \frac{7y}{2} = 21 \Rightarrow y = \frac{2}{7} \times 21 = 6$.

Thus, $x = \frac{5}{2} \times 6 = 15$.

Alternatively, we could say that $2x = 5y \Rightarrow \frac{x}{y} = \frac{5}{2} \Rightarrow x = 5 \times k,\ y = 2 \times k$, where k is a constant of proportionality.

Now, $x + y = 21 \Rightarrow 5k + 2k = 21 \Rightarrow 7k = 21 \Rightarrow k = 3$.

Thus, $x = 5 \times k = 5 \times 3 = 15$ and $y = 2 \times k = 2 \times 3 = 6$.

Thus, $\sqrt{x - y} + \left(\frac{x}{y}\right)^2 = \sqrt{15 - 6} + \left(\frac{5}{2}\right)^2 = \sqrt{9} + (2.5)^2 = 3 + 6.25 = 9.25$.

The correct answer is Option C.

34. The total cost of 75 apples and 45 oranges is given as \$12.

Now, that the cost of any combination of apples and oranges cannot be calculated from the information given as we do not know the per piece price of an apple and an orange.

However, if we look carefully, we find that 50 apples $= \frac{2}{3} \times 75$ apples and 30 oranges $= \frac{2}{3} \times 45$ oranges. The factor of $\frac{2}{3}$ is common to both the fruits.

Thus, we need to calculate the cost of $\frac{2}{3}^{\text{rd}}$ of the quantity of the given combination of fruits.

Thus, the required cost of 50 apples and 30 oranges would be $\frac{2}{3}$ of the cost of 75 apples and 45 oranges $= \frac{2}{3} \times \$12 = \$8$.

The correct answer is Option B.

Alternate approach:

Let the cost of an apple be $\$x$ and that of an orange be $\$y$.

Thus, the cost of 75 apples and 45 oranges would be $\$(75x + 45y)$.

So, we have $75x + 45y = 12 \Rightarrow 3 \times (25x + 15y) = 12 \Rightarrow 25x + 15y = 4 \ldots$ (i)

We now need to calculate the cost of 50 apples and 30 oranges i.e. the value of $(50x + 30y)$.

$(50x + 30y)$can be factored as given below.

$50x + 30y = 2 \times (25x + 15y) = 2 \times 4 = 8$ [Plugging in the value of $25x + 15y = 4$ from equation (i)].

35. Let the number of apples be x and the number of mangoes be y.

Since the number of apples was 4 less than twice the number of mangoes, we have

$x = 2y - 4 \ldots$ (i)

Also, we know that if the man had bought 6 more mangoes, the number of apples and mangoes would have been equal. Thus, we have

$y + 6 = x \ldots$ (ii)

Substituting x from (ii) in (i), we get:

$y + 6 = 2y - 4 => y = 10$.

Thus, the man had bought 10 mangoes.

The correct answer is 10.

36. Let us assume that Lucy was traveling at x miles/hour; $x > 120$ miles/hour. Here, x must be greater than 120 as the fine \$90 is greater than \$60, which implies that she must have sped over 120 miles/hr.

Thus, her fine would be calculated as: $\$[60 + 3(x - 120)]$.

Hence, we have

$60 + 3x - 360 = 90$

$=> 3x = 390 => x = 130$.

Thus, Lucy was driving at 130 miles/hour.

The correct answer is Option C.

37. Let y be the number of coins each thief gets.

Then, $x = 7y + 6$.

Since x & 6 are even, it implies that $7y$ must be even and hence, y is even (Odd × Even = Even).

Let the number that is to be added to x be denoted by a.

Thus, we see that $(x + a)$ is divisible by 14, implying that $(x + a)$ is even (An odd number is not divisible by an even number.).

We know that x is even, hence, we deduce that a is also even.

Hence, we can say that the answer is either option C or D as only those are even numbers.

We need to verify which one of the two options is correct.

To verify with $a = 36$:

$x + 36 = 7y + 6 + 36 = 7y + 42$.

Since y is even, $7y$ is a multiple of 14. Also, 42 is a multiple of 14.

Hence, $7y + 42$ is divisible by 14.

Hence, $a = 36$ satisfies the condition given in the problem. Thus, $a = 42$ is obviously not correct.

The correct answer is option C.

Alternate approach:

We know that when the even number x is divided by 7, the remainder comes to 6.

Obviously, when the same even number x is divided by 2×7 i.e. 14, the remainder would still come to 6.

Thus, we need to add such a number to x, which would leave a remainder of $14 - 6 = 8$ when divided by 14 (only then the remainders would add up to $6 + 8 = 14$ and the resulting number would thus be divisible by 14).

We see that 36 is the only number which when divided by 14 leaves a remainder of 8. Hence, 36 must be the correct answer.

38. Let the age of Jane and Amy at present be x and y years, respectively.

Thus, $y = x - 6 \Rightarrow x - y = 6 \ldots$ (i)

Also, five years ago, the ages of Jane and Amy were be $(x - 5)$ and $(y - 5)$ years, respectively.

Thus, we have $x - 5 = 3(y - 5) \Rightarrow x - 3y = -10 \ldots$ (ii)

Subtracting (ii) from (i): $2y = 16 \Rightarrow y = 8$.

Thus, $x = y + 6 = 14$.

Thus, Jane's age at present is 14 years.

Thus, after 10 years, Jane's age would be $14 + 10 = 24$ years.

The correct answer is Option D.

Alternate approach (using a single variable):

Let, five years ago, the age of Amy was y years.

Thus, at that time, Jane's age = $3y$ years.

Thus, the present ages of Jane and Amy are $(3y + 5)$ and $(y + 5)$ years, respectively.

Hence, as per the first condition, we have $y + 5 = (3y + 5) - 6 \Rightarrow 2y = 6 \Rightarrow y = 3$.

Thus, Jane's age five years ago = $3 \times 3 = 9$ years.

Thus, present age of Jane = $9 + 5 = 14$ years.

Thus, after 10 years, Jane's age would be $14 + 10 = 24$ years.

5.1.5 Absolute numbers

39. $|5x| + 30 = 40$

=> $|5x| = 10$

=> $|x| = 10/5 = 2$

=> $|x| = 2$

=> $x = \pm 2$.

The correct answer is Option C.

40. $\left[\frac{-|-25| \times (-|2|)}{(-|-5|)^2}\right]$

$= \left[\frac{-25 \times (-2)}{(-5)^2}\right]$

$= \frac{50}{25} = 2$.

The correct answer is Option B.

41. $\frac{2x}{3y} = 0$ can be reduced to $\frac{x}{y} = 0 \times \frac{3}{2} = 0$.

From $\frac{x}{y} = 0$, we can deduce two results: one, $x = 0$ & two, $y \neq 0$.

Option D is not necessarily be true. Say, for a condition $(x > y)$, for example, $x = 0$ & $y = -3$, $\frac{x}{y} = 0$ can be true; while options B & C MUST be wrong.

The correct answer is Option A.

42. $|3x - 4| = 11$ means that either $3x - 4 = +11$ or $3x - 4 = -11$

If $3x - 4 = +11$, then $3x = 4 + 11 = 15$ => $x = 5$.

And if $3x - 4 = -11$, then $3x = -11 + 4 = -7$ => $x = -7/3$.

The corrects answers are $x = 5$ and $-7/3$.

43. $\frac{|13x - 4|}{5} + 3 = 10$ => $\frac{|13x - 4|}{5} = 7$ => $|13x - 4| = 35$

$|13x - 4| = 35$ means that either $13x - 4 = +35$ or $13x - 4 = -35$

If $13x - 4 = +35$, then $13x = 4 + 35 = 39$ => $x = 3$.

And if $13x - 4 = -35$, then $3x = -35 + 4 = -31$ => $x = -31/3$.

The corrects answers are $x = 3$ and $-31/3$.

5.1.6 Inequalities

44. $\Rightarrow |2x+1| < 7 \begin{cases} 2x+1 < 7 \Rightarrow 2x < 6 \Rightarrow x < 3. \\ \\ 2x+1 > -7 \Rightarrow 2x > -8 \Rightarrow x > -4. \end{cases}$

So, the range is $-4 < x < 3$.

The correct answer is Option D.

45. Since LHS of the inequity $\frac{1}{x-1} > 1$ is greater than 1, $(x-1)$ must be positive.

We can manipulate the inequality as

$\left[\frac{1}{(x-1)} \times (x-1)\right] > \left[1 \times (x-1)\right]$

$\Rightarrow 1 > (x-1)$

$\Rightarrow 2 > x$; since we know that $(x-1)$ is a positive number, we can multiply both the sides of inequity with a positive number; had it been not known we cannot do it, so considering $(x-1)$ as positive, we have $x > 1$; thus, the range of x would be $1 < x < 2$.

The correct answer is Option D.

46. By multiplying the inequality $\frac{-2|x|}{3y} < -1$ with "–", it will transform to $\frac{2|x|}{3y} > 1$; note the sign change of inequality (from < to >). [When we multiply an inequality with "–", the sign changes.]

Since the numerator $2|x|$ is always a positive number and the expression $\frac{2|x|}{3y}$ is greater than 1, and hence positive, $3y$ or y MUST be positive or $y > 0$.

The correct answer is Option C.

Option A need not be true as unlike y, x may assume a negative number while keeping $|x|$ a positive number.

Option D MUST not be true as for say, $(x > y)$, for an example $x = 3$, & $y = 2$, the inequality $2|x| > 3y$ holds true.

Option B need not be true, since there is no reason for $|x|$ to be less than 3, as shown in the above example.

47. This is a 'Could be true' type of questions. Even if the statement is true for only one scenario, it is the correct answer. 'Could be true' type of questions are different from 'Must be true' type of questions, in which the condition given must be true for all the scenarios.

After multiplying the inequality $-3 < -x < -1$ with "$-$", it can be written as $3 > x > 1$; note the sign change.

Let's take each statement one by one.

I. $x^2 < 2x$: For $1 < x < 2$, the inequality $x^2 < 2x$ is true, else not. Say $x = 1/2$, then $(1/2)^2 < 2 \times (1/2) => 1/4 < 1$; so it can be true.

Alternatively, in case of inequality, if x is a positive number which is true in the given inequality ($1 < x < 3$), then we can cancel x from both the sides. So, $x^{\not 2} < 2\not x => x < 2$. We see that $x < 2$ is within the range of $1 < x < 3$, so Statement I is correct.

II. $x^2 = 2x => x = 2; weseethatx = 2$ is within the range of $1 < x < 3$, so Statement II is also correct. We must not even consider $x = 0$ as $x = 0$ is out of range of given constraints $1 < x < 3$.

III. $x^2 > 2x => x^{\not 2} > 2\not x => x > 2$; Similarly, $x > 2$ is within the range of $1 < x < 3$, so Statement III is also correct.

The correct answer is option D.

48. Since in the inequality $a^3bc^2 < 0$, we see that on the LHS side, there is a term c^2; so we cannot conclude whether c is positive/negative as for c being positive/negative, c is always positive.

However, we can sure conclude that $c \neq 0$ as $a^3bc^2 < 0$. There is no such option of $c \neq 0$, so we move ahead.

We can deduce that a^3bc^2 is negative (Less than 0) because of either a^3 or b, thus a^3b must be negative. The sign of a^3 would depend on the sign of a, if a is positive, a^3 is positive, and if a is negative, a^3 is negative.

a^3b can be negative if and only if one of a or b is negative and the other is positive. Option D states this. Option C is incorrect as it is not "MUST be correct option".

The correct answer is Option D.

49. The direction of both the inequalities must match. Currently, the directions of inequalities: $6 > x > 3$ and $-4 < -y < 7$ are opposite. $-4 < -y < 7$ can be written as $4 > y > -7$.

Since we want $3x$, so by multiplying '3' to $6 > x > 3$, we get $18 > 3x > 9$; similarly, by multiplying '2' to $4 > y > -7$, we get $8 > 2y > -14$.

Adding the inequalities, we get...

$(18 > 3x > 9) + (8 > 2y > -14) => 26 > 3x + 2y > -5$.

$3x+2y > -5$ renders the minimum value for $3x+2y = -4$ and $26 > 3x+2y$ renders the maximum value = 25, thus the set of values is $\{25, 24, 23, ..., 0, -1, -2, -3, -4\}$. There are a total of 30 values.

The correct answer is 30.

5.1.7 Graphic equations & Inequalities

50. The given graph is $f(x) = 2x + 5$.

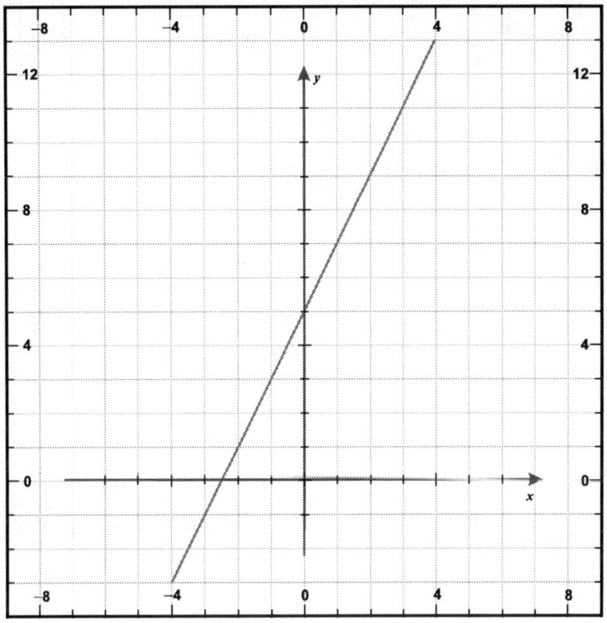

We need to find the graph of $g(x) = 2(x - 1) + 5$.

Here, we see that x has been replaced with $(x - 1)$.

Thus, the graph would shift by one unit right (since x has been reduced by one).

To find the correct graph, let us pick one reference point on the original graph.

Let the point be the one where $f(x)$ intersects the X-axis. The point is $(-2.5, 0)$.

Thus, on shifting, this point would move to $(-2.5, 0 + 1, 0) => (-1.5, 0)$.

Thus, the graph of $g(x)$ should intersect the X-axis at $(-1.5, 0)$.

This is satisfied by the graph shown in option A.

The correct answer is Option A.

51. If we compare the graphs of $f(x)$ and $g(x)$, it seems that $g(x)$ is less steep than $f(x)$ is. Also, we can observe that the point of intersection of $f(x)$ with the Y-axis remains unchanged in $g(x)$ while the point of intersection of $f(x)$ with the X-axis has changed.

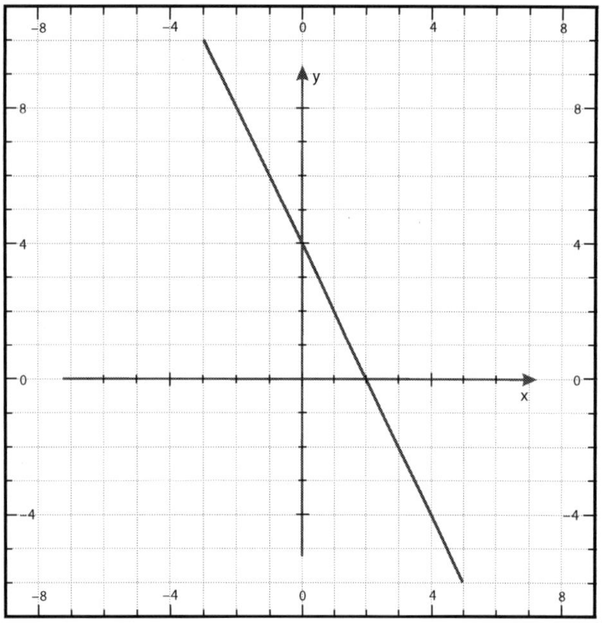

$y = f(x)$

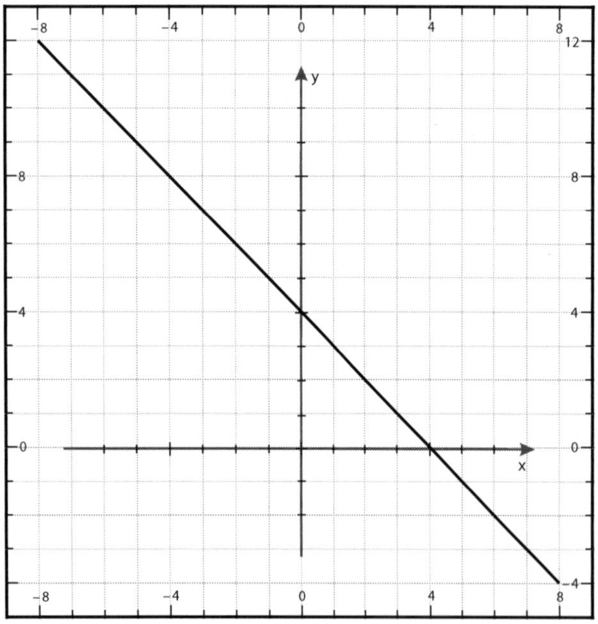

$y = g(x)$

To verify this, let us find the coordinates of the point of intersection of $f(x)$ with the X-axis and the Y-axis. The coordinates of the above points are $(2, 0)$ and $(0, 4)$. Thus, the magnitude of the slope of the line $y = f(x)$ is $\frac{4}{2} = 2$ (magnitude of the slope equals the ratio of the magnitude of the y intercept and the magnitude of the x intercept).

The corresponding points of intersection of $g(x)$ with the X-axis and the Y-axis are $(4, 0)$

and $(0, 4)$. Thus, the magnitude of the slope of the line $y = g(x)$ is $\frac{4}{4} = 1$.

Thus, we see that at the point where the graphs intersect with the X-axis, the value of x has doubled and the magnitude of the slope of the line $y = g(x)$ has halved from $y = f(x)$.

Thus, we can say that $g(x) = f\left(\frac{x}{2}\right)$.

Thus, option D satisfies the above condition.

The correct answer is Option D.

52. On comparing the graphs of $f(x)$ and $g(x)$, it appears that the graph of $f(x)$ has been inverted i.e. reflected about the X-axis to get the graph of $g(x)$.

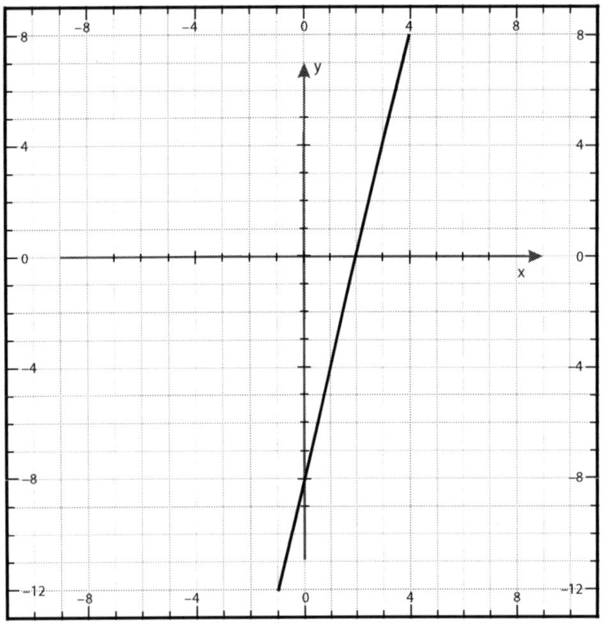

$y = f(x)$

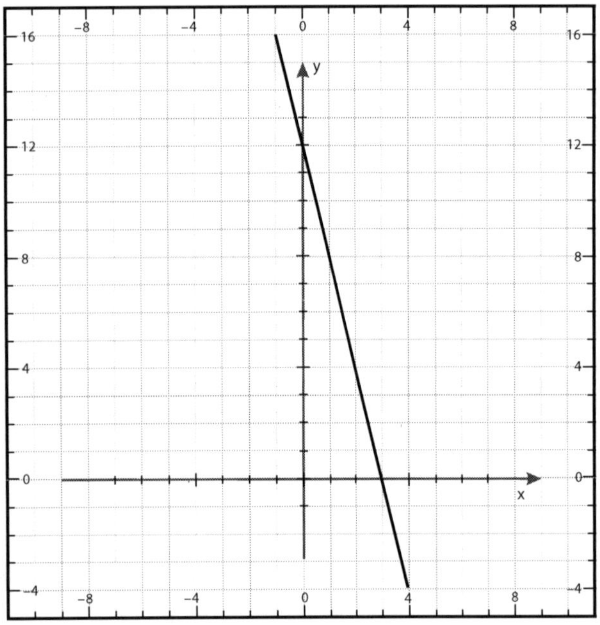

$y = g(x)$

However, simply reflecting the graph of $f(x)$ does not lead to $g(x)$.

Let us pick some reference point in order to compare the graphs of $f(x)$ and $g(x)$.

Let the reference points be the one where $f(x)$intersects the X-axis, which is $(2, 0)$.

In the graph of $g(x)$, the intersection happens at $(3, 0)$.

Thus, we can say that the graph of $f(x)$ has also been shifted one place to the right i.e. x in $f(x)$ has been replaced by $(x - 1)$.

Thus, we can say that there are two changes from $f(x)$ to $g(x)$.

First, we have reflected the graph of $f(x)$ about the X-axis and then shifted it one unit to the right.

Thus, we can say that $g(x) = -f(x-1)$.

Thus, option B satisfies the above condition.

The correct answer is Option B.

53. Let us look at the graph of $|x|$:

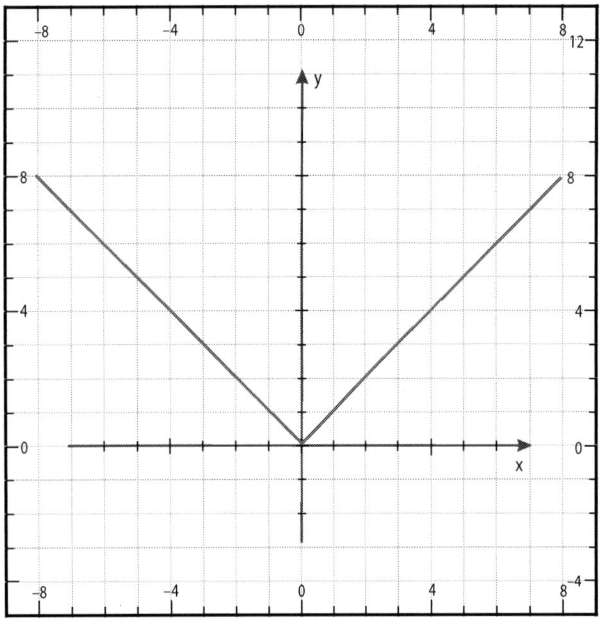

From the above graph of $|x|$, we will try to derive the graph of $4|x-2|-3$.

Let us now look at $|x-2|$.

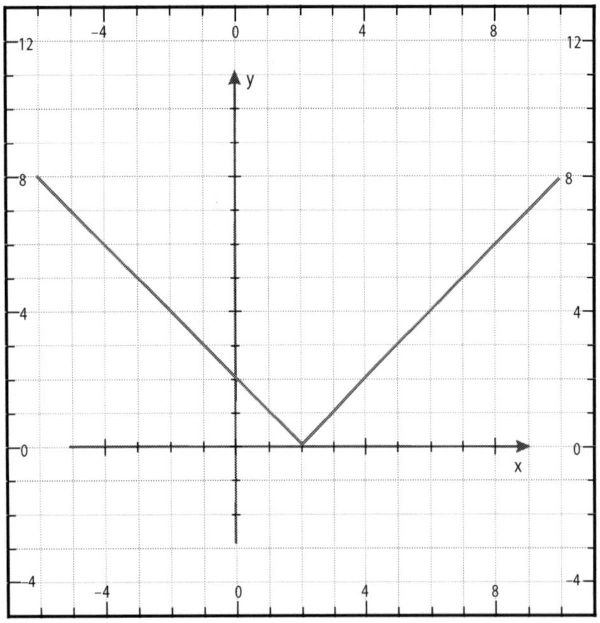

This will be the graph of $|x|$ shifted by 2 units right (since x has been replaced by $(x-2)$).

Let us now look at $4|x-2|$.

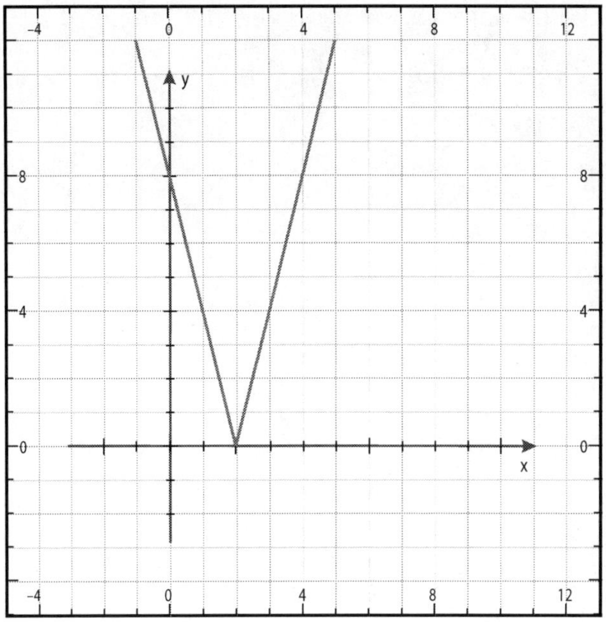

This will stretch the graph of $|x-2|$ vertically making it steeper.

Now, finally we look at the graph of $4|x-2|-3$.

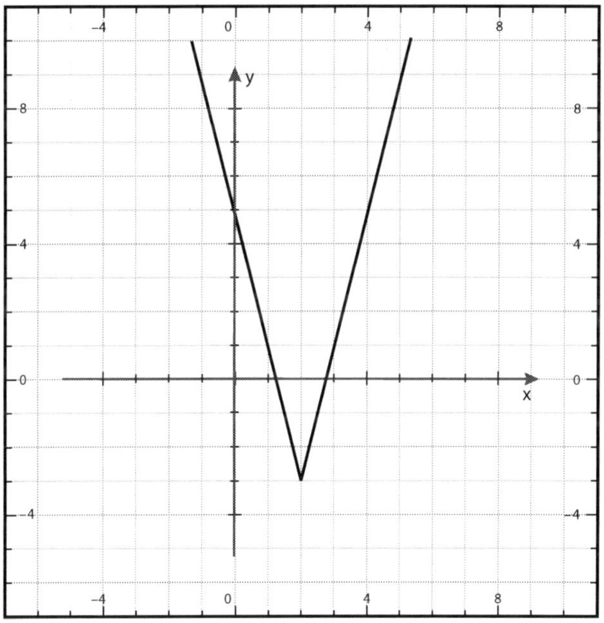

This will shift the graph of $4|x-2|$ down by 3 units (since we have subtracted 3 from the function).

Thus, we have arrived at the graph of $4|x-2|-3$.

However, we need to find the region denoted by $f(x) \geq 4|x-2|-3$.

We can see that all the points that lie inside the V created by the graph have y values which are higher than the values given by $4|x-2|-3$.

To verify the above statement, we pick any point in the V region, say the point $(2,0)$.

Let us calculate the value of $4|x-2|-3$ at the point $x=2$. It gives us $4|2-2|-3=-3$, i.e. the value of the lowest point of the graph.

It is clear that the y value of the point $(2,0)$ i.e. $0>-3$.

Thus, the point $(2,0)$ satisfies the region $f(x) \geq 4|x-2|-3$.

Thus, the region denoted by $f(x) \geq 4|x-2|-3$ will be the shaded region as shown in option A.

The correct answer is option A.

5.1.8 Functions

54. Plug in the value of $x = -2$ in the function,

we get $f(-2) = 3 \times (-2)^3 - 2 \times (-2)^2 + 2 \times (-2) + 2 = -24 - 8 - 4 + 2 = -34.$

The correct answer is Option A.

55. $g(-1/t) = 2(-1/t)^3 - 3(-1/t)^2 + \frac{3}{(-1/t)^2} + \frac{2}{(-1/t)^3}$

$= -\frac{2}{t^3} - \frac{3}{t^2} + 3t^2 - 2t^3 = -(\frac{2}{t^3} + \frac{3}{t^2} - 3t^2 + 2t^3) = -g(t).$

so $g(-\frac{1}{t}) = -g(t).$

The correct answer is Option A.

56. Since the table provides us four values of $f(x)$ for x, we must check each option one by one to see which one represents the given values.

Let us plug-in values of x in each option and check the value for $f(x)$; if all the 4 values for x matched with the respective table values of $f(x)$, it is the correct option.

(A) $f(x) = -4x^2 + 3x + 3$

For $x = 0, f(0) = -4 \times (0)^2 + 3 \times 0 + 3 = 3$; since the value of $f(x)$ matches with the its table value, we must plug-in the next value of x.

For $x = 1, f(1) = -4 \times (1)^2 + 3 \times 1 + 3 = -4 + 3 + 3 = 2$; since the value of $f(x)$ matches with the its table value, we must plug-in the next value of x.

For $x = -1, f(-1) = -4 \times (-1)^2 + 3 \times (-1) + 3 = -4 - 3 + 3 = -4$; since the value of $f(x)$ does NOT match with the its table value, we will not go further as this cannot be an answer. Let's try Option B.

(B) $f(x) = -5x^2 + 4x + 3$

For $x = 0, f(0) = -5 \times (0)^2 + 4 \times 0 + 3 = 3$; since the value of $f(x)$ matches with the its table value, we must plug-in the next value of x.

For $x = 1, f(1) = -5 \times (1)^2 + 4 \times 1 + 3 = -5 + 4 + 3 = 2$; since the value of $f(x)$ matches with the its table value, we must plug-in the next value of x.

For $x = -1, f(-1) = -5 \times (-1)^2 + 4 \times (-1) + 3 = -5 - 4 + 3 = -6$; since the value of $f(x)$ does NOT match with the its table value, we will not go further as this cannot be an answer. Let's try Option C.

(C) $f(x) = 3x^2 - 4x + 3$

For $x = 0, f(0) = 3 \times (0)^2 - 4 \times 0 + 3 = 3$; since the value of $f(x)$ matches with the its table value, we must plug-in the next value of x.

For $x = 1, f(1) = 3\times(1)^2 - 4\times1+3 = 3-4+3 = 2$; since the value of $f(x)$ matches with the its table value, we must plug-in the next value of x.

For $x = -1, f(-1) = 3 \times (-1)^2 - 4 \times (-1) + 3 = 3 + 4 + 3 = 10$; since the value of $f(x)$ matches with the its table value, we must plug-in the next value of x.

For $x = -2, f(-2) = 3 \times (-2)^2 - 4 \times (-2) + 3 = 12 + 8 + 3 = 23$;

Since the values of $f(x)$ matches with the its table values, Option C is the correct answer. We must not check Option D now.

The correct answer is Option C.

57. Since $f(x) = 3x - 4$, hence $f(3x - 4) = 3.(3x - 4) - 4 = 9x - 12 - 4 = 9x - 16$: replacing '$x$' with $3x - 4$.

Now, $5f(x) - 3 = f(3x - 4) => 5.(3x - 4) - 3 = 9x - 16$; substituting the values of $f(x) = 3x - 4$ and $f(3x - 4) = 9x - 16$.

We get, $15x - 20 - 3 = 9x - 16 => 15x - 9x = 23 - 16 => 6x = 7 => x = 7/6$.

The correct answer is $x = 7/6$ or 1.67.

58. $[49] = -|\sqrt{49}| = -7$; as 49 is a positive and an odd integer.

$[-4] = -|(-4)^2| = -|16| = -16$; as -4 is a negative and an even integer.

$[-49] = (-49)^2 = 49^2$; as -49 is a negative and an odd integer.

$[4] = -\sqrt{4} = -2$; as 4 is a positive and an even integer.

So, $\frac{[49]^4\times[-4]}{[-49]\times[4]} = \frac{(-7)^4\times(-16)}{49^2\times(-2)} = \frac{\cancel{49^2}\times\cancel{-16}\,8}{\cancel{49^2}\times\cancel{-2}} = 8$

The correct answer is 8.

59. Set of integers greater than or equal to -2.2 is $\{-2, -1, 0, ...\}$;
among these, the smallest is -2, so, $[-2.2] = -2$.

Similarly, set of integers greater than or equal to 3 is $\{3, 4, 5, ...\}$;
among these, the smallest is 3, so, $[3] = 3$.

Similarly, set of integers greater than or equal to 5.2 are $\{6, 7, 8, ...\}$;
among these, the smallest is 6, so, $[5.2] = 6$.

So, $[-2.2] + [3] + [5.2] = -2 + 3 + 6 = 7$.

The correct answer is Option C.

60. $a@c = \frac{(a + c)}{(a - c)} = 2$

$=> a + c = 2 \times (a - c)$

=> $a + c = 2a - 2c$ => $c = a/3$.

The correct answer is Option C.

5.2 Problem Solving & Data Analysis

5.2.1 Time & Work

61. Since Amy takes 12 hours to paint the entire wall, in 4 hours she had painted only $\frac{4}{12} = \frac{1}{3}^{\text{rd}}$ of the wall.

Thus, Bob has to paint the remaining $\left(1 - \frac{1}{3}\right) = \frac{2}{3}^{\text{rd}}$ of the wall.

We know that Bob takes 18 hours to paint the entire wall.

Thus, to paint $\frac{2}{3}^{\text{rd}}$ of the wall, he would take $\frac{2}{3} \times 18 = 12$ hours.

Hence, it took $(4 + 12) = 16$ hours for them to paint the entire wall.

The correct answer is Option B.

62. John had already completed 12 days of work.

So, the work left (to be done by Matt) is equivalent to 8 days work of John. Note that it is 8 days of work at John's rate.

Hence, the percentage by which the work required to be done by Matt is less than the work already done by John = $\frac{12-8}{12} = \frac{4}{12} \times 100\% = 33.33\% = 33.3\%$ (rounded off)

Hence, the answer is 33.3.

.

63. Let the total work be one unit.

Thus, Joe takes 6 hours to complete 1 unit work. Hence, in one hour, he does $\frac{1}{6}^{\text{th}}$ part of the work.

Also, Joe and Jack together take 4 hours to complete the work.

Hence, in one hour, they do $\frac{1}{4}^{\text{th}}$ part of the work.

Thus, in one hour, Jack alone would do $\left(\frac{1}{4}-\frac{1}{6}\right) = \left(\frac{1}{12}\right)^{\text{th}}$ part of the work.

Hence, to complete the entire work, time taken by Jack = $\frac{1}{\frac{1}{12}} = 12$ hours.

The correct answer is Option D.

Alternate approach:

Joe takes 6 hours, while Joe and Jack together take 4 hours to water the plants.

Hence, let us take the total work to be the LCM of 6 and 4 i.e. 12 units.

Thus, in one hour, Joe does $\frac{12}{6} = 2$ units of work.

Also, in one hour, Joe and Jack together do $\frac{12}{4} = 3$ units of work.

Hence, we can conclude that Jack does $3 - 2 = 1$ unit of work in one hour.

Hence, to complete 12 units of work, Jack would take $\frac{12}{1} = 12$ hours.

5.2.2 Mixtures

64. Given that the concentration of wine is 15%.

To get 15 ml. of wine, we need 100 ml. of the mixture.

To get 27 ml. of wine, we need: $\frac{100}{15} \times 27 = 180$ ml. of mixture.

In the question, the data regarding total volume of mixture, 50 ml. is unnecessary (redundant).

The correct answer is Option D.

65. We know that 50 grams of the metal makes 15% of the total yield.

Thus, 1 gram of the metal makes $= \frac{15}{50}\%$ of the total yield.

Thus, 27 grams of the metal would make $= \frac{15}{50} \times 27 = 8.1\%$ of the yield.

The correct answer is Option C.

66. Given that the ratio is 1 : 3, let the quantity of milk and water in the mixture be x and $3x$, respectively.

Now, 60 liters of milk was added.

So, the quantity of milk in the mixture changed to $(x + 60)$ liters.

However, the quantity of water remained unchanged at $3x$ liters.

Thus, we have $\frac{x+60}{3x} = \frac{3}{5}$ => $5x + 300 = 9x$ => $4x = 300$.

Thus, the total volume of the initial mixture $= x + 3x = 4x = 300$ liters.

The correct answer is Option D.

Alternate approach:

Milk	:	Water
1	:	3
3	:	5

We can solve the question mentally; since no water was added, the quantity of water must remain same. However, in the above representation, we see that the scaled quantity of water has changed from 3 to 5.

To make these same, we multiply 5 with the first ratio and 3 with the second ratio. Thus, we have

Milk	:	Water
$1 \times 5 = 5$	:	$3 \times 5 = 15$
$3 \times 3 = 9$	:	$5 \times 3 = 15$

Now, we see that scaled quantity of milk has changed from 5 to 9 i.e. increased by 4 units. But, we know that these 4 units are equivalent to 60 liters (water added). Thus, 1 such unit is equivalent to 15 liters.

Now, total volume on ratio scale of initial mixture is $5 + 15 = 20$ units.

Thus, 20 units are equivalent to $20 \times 15 = 300$ liters.

67. Let us assume that the total final mixture be 100 units (since this is a percentage based question, the choice of the initial value would not affect the final answer. We choose 100 as it is a number convenient to work with).

In the new mixture formed with water, water forms 60% of the mixture and hence, the energy drink concentrate forms $(100 - 60) = 40\%$ of the mixture i.e. 40% of 100 units = 40 units.

The original concentrate had 20% minerals, i.e. in 100 units of the energy drink concentrate, minerals constituted 20% of the whole.

Thus, in 40 units of the concentrate, minerals constitute 20% of 40 = 8 units.

Thus, percentage of minerals in the new mixture = $\frac{8}{100} \times 100 = 8\%$.

The correct answer is Option B.

68. Let us assume that the total number of customers be 100 (since this is a percentage based question, the choice of the initial value would not affect the final answer. We choose 100 as it is a number convenient to work with).

Thus, 40% of 100 = 40 customers prefer sedans.

Among the customers who prefer sedans, 55% preferred Honda sedan.

Hence, number of customer who prefer Honda sedan = 55% of 40 = $\frac{55}{100} \times 40 = 22$.

Thus, percentage of total customers who prefer Honda sedan = $\frac{22}{100} \times 100 = 22\%$.

Hence, according to the survey, Alfred should have 22% of his cars as Honda sedans.

The correct answer is Option B.

5.2.3 Word Problems

69. Charge for the first 5 lbs of luggage = $\$x$.

Total charge paid by the customer = $\$a$.

Hence, extra charges paid = $\$(a - x)$.

Charge per lb of extra luggage = $\$y$.

Hence, weight of extra luggage = $\frac{a-x}{y}$ lbs.

Hence, total weight of luggage carried by the customer = $5 + \frac{a-x}{y}$ lbs.

The correct answer is Option B.

70. The initial salary was $\$x$.

As a result of a 25% increase, the salary will increase by 25% or $\frac{1}{4}$ times the initial value i.e. it will become $\$(x + \frac{1}{4} \times x) = \$\frac{5x}{4}$.

It is given that the salary after the increase = \$400.

Thus, $\frac{5x}{4} = 400 => x = 400 \times \frac{4}{5} = 320$.

Hence, $x - 100 = 320 - 100 = 220$. Note that the question does not ask for the value of x; it asks for the value of $(x - 100)$. Beware!

The correct answer is 220.

71. The firm manufactures screws, rivets and nuts for 60 minutes, 60 minutes and 120 minutes, respectively or $60 \times 60 = 3,600$ seconds, $60 \times 60 = 3,600$ seconds, $120 \times 60 = 7,200$ seconds, respectively.

Number of screws manufactured = $\frac{3,600}{10} = 360$

Number of rivets manufactured = $\frac{3,600}{20} = 180$

Number of nuts manufactured = $\frac{7,200}{80} = 90$

Thus, total items manufactured = $360 + 180 + 90 = 630$.

Thus, nuts expressed as a percentage of the total items
$= \frac{90}{630} \times 100 = \frac{1}{7} \times 100 = 14.28\% =\approx 14.3\%$.

The correct answer is 14.3.

Alternate approach:

Ratio of the times taken to manufacture a screw, a rivet & a nut = $10 : 20 : 80 = 1 : 2 : 8$.

Thus, ratio of the number of screws, rivets & nuts manufactured in the same duration $= \frac{1}{1} : \frac{1}{2} : \frac{1}{8} = 8 : 4 : 1$.

However, the duration (no. of hours) for which number of screws, rivets & nuts are manufactured

$= 1 : 1 : 2.$

Thus, ratio of the number of screws, rivets & nuts manufactured

$= 8 \times 1 : 4 \times 1 : 1 \times 2 = 8 : 4 : 2 = 4 : 2 : 1.$

Thus, number of nuts expressed as a percentage of the total items

$= \frac{1}{4+2+1} \times 100 = \frac{1}{7} \times 100 = 14.28\% = \approx 14.3\%.$

72. Total number of items $= 22 + 37 + 13 = 72.$

Since each dealer has the same number of items, each must have $= \frac{72}{3} = 24$ items.

Now, according to the question, each dealer must get at least 2 items of each product.

Hence, to maximize the number of items of R with the dealer who gets the maximum units of R (let's call it the third dealer), let us give 2 items of R to each of the other two dealers. Thus, we are left with $= 13 - 2 - 2 = 9$ items of R.

All the above 9 items of R can be given to the third dealer. Now the third dealer will get $24 - 9 = 15$ items of product P and Q. Since P and Q, together are 59, which is far greater than 15, this scenario is possible.

Thus, the maximum number of items of R that can be present with any dealer is 9.

The correct answer is Option B.

73. The bus makes a total of $6 \times 3 = 18$ trips in a week.

Thus, total distance travelled by the bus in a week $= (18 \times d)$ miles.

Thus, total fuel cost for the week $= \$(18 \times d) \times n = \$(18 \times d \times n)$

We know that the total operating cost for the week is $\$c$.

c = Total fuel cost + Maintenance cost

Hence, the maintenance cost for the week $= \$(c - 18dn)$

This is the maintenance cost over 6 days (since the bus operates on 6 days of the week).

Thus, the daily maintenance cost $= \$\frac{(c-18dn)}{6}$.

The correct answer is Option C.

74. Total number of items $= 4,500.$

Number of defective $= 30\%$ of $4,500 = \frac{30}{100} \times 4,500 = 1,350.$

Number of items rectified $= 80\%$ of $1,350 = \frac{80}{100} \times 1,350 = 1,080.$

Number of items scrapped $= 1,350 - 1,080 = 270.$

Hence, number of items rectified after rework are more than those which are scrapped by $= 1,080 - 270 = 810$.

The correct answer is Option B.

Alternate approach:

Percentage of defective items = 30%.

Thus, as a percentage of the total items, the number of defective items rectified after rework = 80% of 30% $= \frac{80}{100} \times 30\% = 24\%$.

Again, as a percentage of the total items, the number of defective items scrapped

$= (100 - 80)\%$ of $30\% = \frac{20}{100} \times 30\% = 6\%$.

Hence, defective items rectified after rework is more than that scrapped items by

$= (24 - 6)\,\%$ of the total items = 18% of the total items $= \frac{18}{100} \times 4,500 = 810$.

75. Ice forms at the rate of 2 inches in every 40 minutes or $\frac{40}{60} = \frac{2}{3}$ hours.

Thus, the rate of ice formation $= \frac{2}{\frac{2}{3}} = 3$ inch per hour

Thus, in 4 hours, total thickness of ice formed $= 3 \times 4 = 12$ inches.

The freezer was switched off for 2 hours and the ice melted at the rate of 0.5 inches per hour.

Thus, total amount of ice melted $= 2 \times 0.5 = 1$ inch.

Thus, ice left $= 12 - 1 = 11$ inches.

Finally, the freezer was put on for another 2 hours

Ice formed during these 2 hours $= 3 \times 2 = 6$ inches.

Thus, total thickness of ice now $= 11 + 6 = 17$ inches.

The correct answer is Option C.

76. Fraction of the total books given to children aged less than 8 years $= \frac{3}{11}$.

Hence, fraction of books remaining $= 1 - \frac{3}{11} = \frac{8}{11}$.

Fraction of the remaining books given to children aged between 8 and 12 years $= \frac{3}{4}$ part of remaining books $\left(\frac{8}{11}\right) = \frac{3}{4} \times \frac{8}{11} = \frac{6}{11}$.

Hence, the fraction of books remaining $= \frac{8}{11} - \frac{6}{11} = \frac{2}{11}$ = books given to children older than 12 years.

Hence, fraction of books given to children older than 12 years $= \frac{2}{11}$.

The correct answer is 2/11 or 0.18.

77. The percentage reduction in the enrollment of students from 2014 to 2015 is calculated as:

$$\% \text{ Reduction} = \frac{\text{Reduction}}{\text{Value in 2014}} \times 100 = \frac{235 - 200}{235} \times 100 = 14.89\% \ =\approx 14.9\%$$

The correct answer is 14.9.

78. To manufacture the item, Process A requires 540 components.

These components can be manufactured @60 components per 25 minutes.

Thus, time taken to manufacture 540 components = $\frac{540}{60} \times 25 = 9 \times 25 = 225$ minutes.

To manufacture the same item, Process B requires 300 components.

These components can be manufactured @50 components per 15 minutes.

Thus, time taken to manufacture 300 components = $\frac{300}{50} \times 15 = 6 \times 15 = 90$ minutes.

Thus, percentage reduction in time by replacing Process A with Process B

$= \frac{\text{Reduction in time}}{\text{Time taken by Process A}} \times 100$

$= \frac{225-90}{225} \times 100\% = \frac{135}{225} \times 100 = 60\%.$

The correct answer is 60.

79. Number of people who were eligible to vote in 2010 = 192 million.

This number increased by 10.8 million in 2011.

Hence, the number of people who were eligible to vote in 2011 = 192 + 10.8 = 202.8 million.

Number of people who voted in 2010 = 54 million.

This number increased by 2.4 million in 2011.

Hence, the number of people who voted in 2011 = 54 + 2.4 = 56.4 million.

Hence, the percentage of people who voted of people who were eligible to vote in 2011

$= \frac{56.4}{202.8} \times 100 = 27.81\%$

The correct answer is Option B.

80. Let us assume that the total number of students registering for the AST be 100 (since this is a percentage based question, the assumed initial value will have no effect on the answer).

Number of students appearing for the test = 80% of 100 = 80.

Thus, number of students who do not appear after registering = 100 − 80 = 20.

Of the 80 students who appear for the test, only 75% clear the test.

Thus, number of students who clear the test = 75% of 80 = $\frac{75}{100} \times 80 = 60$.

Thus, the percentage by which the number of students who do not appear for the test after registering is less than the number of students who clear the test is given by:

$$\frac{(\text{\# who clear the test}) - (\text{\# who do not appear after registering})}{\text{\# who clear the test}} \times 100 = \frac{60-20}{60} \times 100\% = 66.66\%.$$

The correct answer is Option D.

81. Total number of students registering for the AST = 50,000.

Number of students appearing for the test = 80% of 50,000 = 40,000.

Thus, number of students who do not appear after registering
= 50,000 − 40,000 = 10,000.

Of the 40,000 students who appear for the test, only 75% clear the test.

Thus, number of students who clear the test = 75% of 40,000 = $\frac{75}{100} \times 40{,}000 = 30{,}000$.

Thus, the difference between the number of students who do not appear for the test after registering and the number of students who clear the test = 30,000 − 10,000 = 20,000

The correct answer is Option B.

82. The man purchased goods at the rate of 12 for \$5 and sold them at the rate of 10 for \$6.

Let us assume that the number of goods purchased = LCM of 12 and 10 = 60 (since this is a percentage based question, the assumed initial value will have no effect on the answer. So, we chose a number i.e. 60 which is divisible by both 12 and 10 to make our calculations easier).

Thus, the total cost of 60 goods = $\$\frac{5}{12} \times 60 = \$25$.

The total sales proceeds generated by selling all the goods = $\$\frac{6}{10} \times 60 = \$36$.

Thus, profit made on selling all the goods = \$ (36 − 25) = \$11.

Hence, percentage profit generated = $\frac{\text{profit}}{\text{total cost}} \times 100 = \frac{11}{25} \times 100 = 44\%$.

The correct answer is Option D.

83. Cost of a 10 second advertisement during prime-time = \$12,000.

The cost of the same advertisement early morning is 30% lower than that during prime-time.

Thus, cost of a 10 second advertisement during early morning
= \$(12,000 − 30% of 12,000) = \$8,400.

Since the advertisement is of 30 seconds duration, we first calculate the cost for the given 10 second duration advertisement and then multiply the answer by $\frac{30}{10}$, i.e., 3.

Hence, total cost of running a 10 second advertisement once during prime-time and twice early morning $= \$(12,000 + 8,400 \times 2) = \$28,800$.

Hence, total cost of running a 30 second advertisement once during prime-time and twice early morning

$=\$28,800 \times 3 = \$86,400 = \$86.4$ thousand dollars.

The correct answer is Option C.

84. We know that the average speed $= \dfrac{\text{Total distance}}{\text{Total time}}$

In the problem, the sprinter follows the exact same route during both ways and hence, her distance was identical in either case.

Let us assume that the distance travelled be the product $s(2s + 1)$.

Since speed is a ratio of distance and time, the ratio of distance and time would remain the same irrespective of the value assumed for the distance. Here we assumed the distance to be the product (or the LCM) of the speeds since this would help us in our calculations.

Since the speed up the slope was s, time taken to cover the distance $= \frac{s(2s+1)}{s} = (2s + 1)$.

Again, since the speed down the slope was $(2s + 1)$, time taken to cover the distance $= \frac{s(2s+1)}{2s+1} = s$.

Thus, total time taken $= (2s + 1) + s = (3s + 1)$.

Total distance covered going up the slope and then down $= 2s(2s + 1)$

Hence, her average speed for the entire journey $= \frac{\text{Total distance}}{\text{Total time}} = \frac{2s(2s+1)}{3s+1}$.

The correct answer is option D.

85. Let us assume that the time taken to go up the slope be t minutes.

Since the total time is 6 minutes, so time taken to come down the slope $= (6 - t)$ minutes.

Speed of the sprinter up the slope = 3.6 feet/second.

Hence, distance covered up the slope $= 3.6t \times 60 = 216t$ feet (The speed is in feet/second while time was in minutes. So, we multiplied 60 to convert the time from minutes to seconds).

Similarly, speed of the sprinter down the slope = 6 feet/s.

Hence, distance covered up the slope $= 6(6 - t) \times 60 = 2,160 - 360t$ feet (The speed is in feet/second while time was in minutes. So, we multiplied 60 to convert the time from minutes to seconds).

Since distance covered both ways is the same, we have

$216t = 2,160 - 360t \Rightarrow 576t = 2,160 \Rightarrow t = \frac{2160}{576} = 3.75$.

Thus, time taken to go up the slope = 3.75 minutes.

Thus, one-way distance = $3.6t \times 60 = 3.6 \times 3.75 \times 60 = 810$ feet.

The correct answer is Option B.

86. We know that the average speed = $\dfrac{\text{Total distance}}{\text{Total time}}$

For the first part of the journey:

Distance = d miles.

Speed = 60 miles/hr.

Hence, time taken = $\frac{d}{60}$ hours.

For the second part of the journey:

Distance = 60 miles.

Time taken = 6 hours.

Hence, total distance for the entire journey = $(d + 60)$miles.

Total time taken for the journey = $\left(\frac{d}{60} + 6\right)$ hours.

Hence, average speed = $\frac{\text{Total distance}}{\text{Total time}} = \frac{d+60}{\frac{d}{60}+6} = \frac{60(d+60)}{d+360}$ miles/hr.

The correct answer is Option A.

87. We need to find the total number of students in the class.

The main idea is that students who play both games are already counted among the students who play football as well as among the students who play rugby.

The best idea is to break the students in three categories: those who play only football, those who play only rugby and those who play both games.

The number of students who play only football is the difference between the number of students who play football and the number of students who play both games
$= 40 - 10 = 30$.

The number of students who play only rugby is the difference between the number of students who play rugby and the number of students who play both games
$= 30 - 10 = 20$.

Thus, total number of students who play some game is the sum of the students who play only football, who play only rugby and who play both games
$= 30 + 20 + 10 = 60$.

Now, we know that of the total number of students, 20% do not play any game. Thus, the number of students who play some game is $(100 - 20) = 80\%$ of the total students.

Thus, we have 80% of total students = 60 => total students = $\frac{100}{80} \times 60 = 75$.

Hence, total students in the class = 75.

88. Total duration of peak hours (from 9:00 am to 1:00 pm) = 4 hours = $4 \times 60 = 240$ minutes.

Since cars arrive every p minutes, number of cars that arrive for fueling during this time $= \frac{240}{p}$ cars.

Total duration of non-peak hours (from 3:00 pm to 6:00 pm) = 3 hours = $3 \times 60 = 180$ minutes.

Since cars arrive every q minutes, number of cars that arrive for fueling during this time $= \frac{180}{q}$ cars.

Thus, total number of cars that arrived for fueling = $\left(\frac{240}{p} + \frac{180}{q}\right)$.

Total time duration = 4 + 3 = 7 hours = $7 \times 60 = 420$ minutes.

Thus, average number of cars arriving per minute

$= \frac{\left(\frac{240}{p} + \frac{180}{q}\right)}{420} = \frac{240q + 180p}{420pq} = \frac{60(4q + 3p)}{420pq} = \frac{4q + 3p}{7pq}$.

The correct answer is Option C.

89. We have the following information:

#Apples = $\frac{1}{3} \times$ #Mangoes => #Mangoes = $3 \times$ #Apples... (i)

#Guavas = $\frac{1}{4} \times$ #Oranges => #Oranges = $4 \times$ #Guavas ... (ii)

#Mangoes = $\frac{1}{2} \times$ #Guavas => #Guavas = $2 \times$ #Mangoes ... (iii)

Since this is a ratio question, the choice of the initial number we assume will not affect the final answer. We need to choose a convenient value for the number of apples. Then, we can find out the number of fruits of all the other types using the relations above.

Let us assume #Apples = 1.

Thus, from (i): #Mangoes = $3 \times 1 = 3$.

From (iii): #Guavas = $2 \times 3 = 6$.

From (ii): #Oranges = $4 \times 6 = 24$.

Thus, total number of fruits = $1 + 3 + 6 + 24 = 34$.

Thus, fraction of total fruits that are Mangoes = $\frac{3}{34} = 0.088 =\approx 0.09$.

The correct answer is 3/34 or approximately 0.09.

90. We have the following information:

#CDs = $\frac{1}{3}$ × #Pen-drives => #Pen-drives = 3 × #CDs . . . (i)

#DVDs = $\frac{1}{4}$ × #Pen-drives => #Pen-drives = 4 × #DVDs . . . (ii)

Since this is a ratio questions, the choice of the initial number we assume will not affect the final answer. We need to choose a convenient value for the number of Pen-drives. Then, we can find out the number of items of all the other types using the relations above.

We can see that number of Pen-drives is thrice the number of CDs as well as four times the number of DVDs. So, we can say that the number of Pen-drives is a multiple of both 3 and 4 and hence, a multiple of their LCM. The LCM of 3 and 4 is 12.

So, let us assume #Pen-drives = 12.

Thus, from (i): #CDs = $\frac{1}{3} \times 12 = 4$.

From (ii): #DVDs = $\frac{1}{4} \times 12 = 3$.

Thus, total number of items = $12 + 4 + 3 = 19$.

Thus, fraction of total items that are DVDs = $\frac{3}{19} = 0.157 =\approx 0.16$.

The correct answer is 3/19 or approximately 0.16.

91. The largest number of boxes that can fit is obtained by placing the lowest dimension of the box, 2 meters, against the largest dimension of a container, 20 meters.

So, the length of the container can allow for $\frac{20}{2} = 10$ boxes, and the width of the container allows for $\frac{6}{3} = 2$ boxes to be fit, and the height of the container fits just $\frac{4}{4} = 1$ box.

Hence, the total number of boxes that fit in a single container is $10 \times 2 \times 1 = 20$.

The correct answer is 20.

92. If the device was losing 10 centimeters for every 15 meters, then we need to know how many 15-meter intervals are there in 1,800 meters. Conveniently, $1,800/15 = 120$. So, at the end of the bridge, the device was off by 1,200 centimeters, or 12 meters. Since the device underestimated the distance, the length of the bridge, according to the device, was $1,800 - 12 = 1,788$ meters.

The correct answer is Option D.

93. Suppose it costs 1 dollar to hire an MA student, so it costs 2 dollars to hire an MBA. Then the total cost of hiring is $2 \times 140 + 40 = 320$ dollars, including the cost of hiring of the students with MA degrees is 40 dollars.

So, the cost of hiring all MA's is $\frac{40}{320} \times 100\% = 12.5\%$ of the total cost of hiring

The correct answer is 12.5.

94. The average rate of arrival during the day is the ratio of "the total number of customers who arrive at the bank during the day" to "the total number of hours: $t_1 + t_2$".

The rate of arrival during the morning is given in customers per unit of time. So, multiplying the rate by hours will yield the number of customers arriving to the bank in the morning. The same is true for the afternoon.

Hence, the total number of customers who arrive to the bank during the day is $(r_1 \times t_1 + r_2 \times t_2)$, and the average rate of arrival for the day is $\frac{r_1t_1+r_2t_2}{t_1+t_2}$.

The correct answer is Option C.

5.2.4 Profit & Loss

95. The key to this question is to know that, by definition, the percent change in price is the difference of old and new prices divided by the old price, or, mnemonically, it is 'new minus old divided by old'. So, given the numbers of the problem, the percent change is: $\frac{90-120}{120} \times 100 = -\frac{30}{120} \times 100 = -25\%$

Since the selling price is less than the cost price, the dealer incurred a loss which is shown by the "−" sign in the answer. Since the question specifically asks for the loss, we do not need to append the "−" with the answer.

The correct answer is Option B.

96. The cost price for the distributor is \$120.

His selling price is 10% higher than the cost i.e. $\$(120 + 120 \times \frac{10}{100}) = \$132$.

However, he wishes to get this price of \$132 after giving a 20% discount on the list price of the product.

Thus, if the list price be $\$x$, then he gave $\$\frac{x}{5}$ off on the price (20% = 1/5).

Thus, his selling price would be $\$(x - \frac{x}{5}) = \$\frac{4x}{5}$.

Hence, $\frac{4x}{5} = 132 => x = 132 \times \frac{5}{4} = 33 \times 5 = 165$.

Hence, the list price for the product must have been \$165.

The correct answer is Option D.

Alternate approach:

We can understand from the question that the list price must be more than the cost price for the distributor.

Thus, Option A cannot be correct.

Again, the list price must be more than the selling price of the product.

His selling price is 10% higher than the cost i.e. $\$(120 + 120 \times \frac{10}{100}) = \$132$.

Thus, options B and C cannot be correct either.

97. We know that the cost of 16 toffees is \$4.

Thus, to make 25% profit, the price at which the man must sell those 16 toffees is $\$(4 + 4 \times \frac{25}{100}) = \$5$.

So, for \$5, the man sells 16 toffees.

Thus, for \$10, the man would sell $\frac{16}{5} \times 10 = 32$ toffees.

The correct answer is Option A.

98. The cost of 50 balloons if purchased in single pieces = $0.10 × 500 = $50.

To purchase the balloons in boxes, one would need to buy $\frac{500}{25}$ = 20 boxes.

The cost of 20 boxes $2 × 20 = $40.

Hence, savings in cost = $50 − $40 = $10.

The correct answer is Option B.

99. The cost of balloons purchased in single pieces = $0.10.

The effective cost of each balloon if purchased in boxes = $\$\frac{2}{25}$ = $0.08.

We see that the cost of balloons is lower if purchased in boxes. Hence, we need to purchase the balloons as many as possible in boxes in order to minimize the cost.

Now, we know that each box has 25 balloons.

Hence, in order to buy 310 balloons, the maximum number of boxes that can be purchased is the quotient when we divide 310 by 25, which is 12 (since 12 boxes would give 12 × 25 = 300 balloons).

The remaining 310 − 300 = 10 balloons would have to be purchased in single pieces.

The cost of 12 boxes = $2×12 = $24 and the cost of 10 single balloons = $0.10×10 = $1.

Hence, the total cost = $24 + $1 = $25.

The correct answer is Option C.

100. Let the cost of one pencil be $\$x$.

Thus, the cost of one pen = $\$3x$ (since a pen costs thrice as much as a pencil).

Hence, the total cost of 10 pens and 10 pencils = $\$(10 \times x + 10 \times 3x) = \$40x$.

Thus, $40x = 20 \Rightarrow x = 0.50$.

Hence, the cost of 5 pencils = $5 × 0.50 = $2.50 = 250 cents.

The correct answer is 250.

101. We know that the cost of one pencil is $0.50.

Thus, the cost of one pen = $3 × 0.50 = $1.50 (since a pen costs thrice as much as a pencil).

Since the boy must purchase 3 pens, expense for the pens = $3 × 1.50 = $4.50.

Amount of money left with the boy = $(20 − 4.50) = $15.50.

Since the cost of a pen is more than that of a pencil, the number of items would be maximized if the boy purchased only pencils with the remaining amount.

Since the cost of each pencil is \$0.50, the number of pencils he can buy $= \$\frac{15.5}{0.50} = 31$.

Hence, the boy can purchase 31 pencils. Hence, the total items he can purchase $= 31 + 3 = 34$.

The correct answer is 34.

Note that '31' as an answer is wrong since the total number of items includes '3 pens' too.

102. We know that the selling price of 7 DVDs equals the cost price of 9 DVDs.

Let us assume that the selling price of 7 DVDs = cost price of 9 DVDs = LCM of 7 and 9 = \$63.

We take the LCM, a convenient number, since we need to take a value divisible by both 7 and 9. We could have taken \$100 as well, but then the calculation would have then been difficult.

Thus, the selling price of 9 DVDs $= \$\frac{63}{7} \times 9 = \$81$.

Thus, the profit made on 9 DVDs $= \$(81 - 63) = \$18$.

Hence, percentage profit

$= \frac{\text{Total profit made in selling 9 DVDs}}{\text{Cost price of 7 DVDs}} \times 100\% = \frac{18}{63} \times 100 = 28.57\% = \approx 28.6\%$

The correct answer is 28.6.

103. We know that the selling price of 7 DVDs equals the cost price of 9 DVDs.

Let us assume that the selling price of 7 DVDs = cost price of 9 DVDs

$= k \times$ (LCM of 7 and 9) $= \$63k$ (where k is a constant of proportionality).

We take the LCM since we need to take a value divisible by both 7 and 9. We could have taken \$100 as well, but then the calculation would have then been difficult.

Thus, the selling price of one DVD $= \$\frac{63k}{7} = \$9k$ and

the cost price of one DVD $= \$\frac{63k}{9} = \$7k$.

Thus, the amount saved by the customer on each DVD $= \$(9k - 7k) = \$2k$.

So, the amount saved by the customer on 10 DVDs $= \$2k \times 10 = \$20k$.

Since the man saves \$30, we have $20k = 30 \Rightarrow k = 1.5$.

Thus, the cost price of one DVD $= \$7 \times 1.5 = \$10.5$.

The correct answer is 10.5.

104. Say, the price of one dinner is \$1, thus for 5 dinners, it would be \$5, but the restaurant will ask you to pay for only 4 dinners or \$4. This way, they get a discount of \$5 – \$4 = \$1 on \$5 bill.

=> % discount = $\frac{1}{5} \times 100\% = 20\%$

Note that the discount would be calculated on \$5 and not on \$4, else the answer would have been 25%, which is wrong.

The correct answer is 20.

105. Since the man will get 3 shirts free, he should buy only 5 shirts.

Say, the price of one shirt is \$1, thus for 5 shirts, it would be \$5, and for 8 shirts, it would be \$8. This way, he gets a discount of \$8 – \$5 = \$3 on \$8 bill.

=> % discount = $\frac{3}{8} \times 100\% = 37.5\%$

Note that the discount would be calculated on \$8 and not on \$5, else the answer would have been 60%, which is wrong.

The correct answer is 37.5.

106. If P is the price of the boat, then $P + 7\%$ of $P = 1.07P = 6,885.45$.

Thus, $P = \frac{6,885.45}{1.07} = \$6,435$

The correct answer is option C.

107. The question asks which of the numbers given in the options cannot be a price of the item when its original price, in whole number, is raised by 10%.

Say the original price of the item is x, a whole number, thus after the raise of 10%, it would be $x + 10\%$ of $x = 1.1x$.

So, of the four options, only three option can be equated to $1.1x$, resulting x a whole number. The option that results x in a decimal number is the correct answer.

Let us each option one by one.

(A) It implies that $1.1x = 5.50 \Rightarrow x = \frac{5.50}{1.1} = 5$, a whole number, thus is can be the resulting price of the item.

(B) It implies that $1.1x = 6.60 \Rightarrow x = \frac{6.60}{1.1} = 6$, a whole number, thus is can be the resulting price of the item.

(C) It implies that $1.1x = 11.70 \Rightarrow x = \frac{11.70}{1.1} = 10.64$, not a whole number, thus is cannot be the resulting price of the item. It is the correct answer.

Since we got the correct answer, there is no need to check option D.

The correct answer is Option C.

108. The listed price of each book =\$60.

Thus, selling price of the first book = \$60.

The customer gets 10% discount on each of the remaining three books.

Thus, discount on each book = $\$(10\% \text{ of } 60) = \$(\frac{10}{100} \times 60) = \$6$.

Thus, selling price of each of the remaining three books = $\$(60 - 6) = \$54$.

Thus, total amount paid by the customer = $\$(60 + 54 \times 3) = \$(60 + 162) = \$222$.

The correct answer is 222.

5.2.5 Interest & Compounding

109. Let us assume that initially there are 100 bacteria (since this is a percentage based question, the initial number of bacteria does not matter and 100 is a convenient number).

So, initially, after one hour, the number of bacteria present = $2 \times 100 = 200$.

So, initially, after two hours, the number of bacteria present would be $2 \times 200 = 400$.

With the introduction of chemical agent, after two hours, the same 100 bacteria would have grown in numbers to $3 \times 100 = 300$. (We need to keep the time period same in both cases in order to compare the growths).

Thus, the agent reduced the growth by $400 - 300 = 100$ bacteria.

Thus, percentage reduction in growth $= \frac{\text{Reduction}}{\text{Initial value}} \times 100 = \frac{100}{400} \times 100\% = 25\%$.

The correct answer is 25.

110. We have to calculate the interest at the end of a year based on three terms of payment.

(A) interest calculated at the end of the year at the rate of 10% per annum

The interest at the end of the year = $\$1,000 \times 10\% = \$100$

(B) interest calculated at the end of every 6 months at the rate of 9% per annum

Since the interest is calculated at the end of every 6 months, it would be calculated two times for a period of 1 year.

The interest at the end of the first 6 months = $\$1,000 \times \frac{9}{2}\% = \$45$ (6 month rate of interest would be $\frac{9}{2}\%$)

Now the sum for the next 6-month would be $\$1,000 + 45 = \$1,045$

The interest at the end of the first 6 months = $\$1,045 \times \frac{9}{2}\% = \$47$

Thus, the total interest = $\$45 + \$47 = \$92 < \$100$ (interest from Option A)

(C) interest calculated at the end of every 3 months at the rate of 8% per annum

Since the interest is calculated at the end of every 3 months, it would be calculated four times for a period of 1 year.

Since calculating interest 4 times would be time-consuming, we apply formula of compounding.

$A = P\left(1 + \frac{r}{100}\right)^n$;

where A = Amount after n periods of interest, P = sum invested, and r = rate of interest

Thus, interest = $A - P$

As per the term of payment, $A = 1,000\left(1 + \frac{2}{100}\right)^4$; (3 month rate of interest would be $\frac{8}{4} = 2\%$; $n = 4$, as the interest would be calculated 4 times in a year)

$A = 1,000\left(1 + \frac{2}{100}\right)^4 = 1,000 \times 1.02^4 = 1,082$

Thus, interest = $1,082 - 1,000 = \$82 < \$100$ (interest from Option A)

Thus, Option A yields the most interest.

The correct answer is Option A.

111. The gain or the loss at the end of the year

= Two half-yearly payments – One annual payment

Annual payment = $20,000 \times 6\frac{1}{4}\% = \frac{20,000 \times 25}{100 \times 4} = \$1,250$

Two half-yearly payments = $2 \times \left[20,000 \times (\frac{6}{2})\% = 20,000 \times 3\% = \frac{20,000 \times 3}{100}\right] = \$1,200$

Thus, loss = $1,250 - 1,200 = \$50$.

Note: The question states that the interest calculation as per the new term is not compounded.

The correct answer is Option B.

112. Let the phone price be x.

First offer:

Effective price of the phone = 20% of x + 5 × 20% of x

=> $0.2x + 5 \times 0.2x = 0.2x + x = 1.2x$.

Effective price of the phone =$1.2x$.

Second offer:

Effective price of the phone = 10% of x + 12 × 10% of x

=> $0.1x + 12 \times 0.1x = 0.1x + 1.2x = 1.3x$.

Effective price of the phone = $1.3x$.

The phone is cheaper in the first offer by

$= \left[\frac{\text{Eff. price from second offer} - \text{Eff. price from first offer}}{\text{Eff. price from first offer}}\right] \times 100\%$

$= \left[\frac{1.3x - 1.2x}{1.3x}\right] \times 100\% = \frac{0.1x}{1.3x} \times 100\% = 7.69\%$

The correct answer is Option A.

113. Let the initial bacteria count be 1. We want the time when the bacteria count would be more than $1.5 \times 1 = 1.50$.

Or, Bacteria count (then) > 1.5× Bacteria count (now)

Given that the count is growing at 10% per minute.

After one minute, the bacteria count would be $1 \times (1 + 10\%) = 1 \times 1.1 = 1.1$.

Similarly, after two minutes, the bacteria count would be $1 \times (1.1)^2 = 1.21 < 1.50$, but it is less than the desired value, 1.5.

Again, after four minutes, the bacteria count would be $1 \times (1.1)^4 = 1.46 < 1.50$, still less than 1.5.

Again, after five minutes, the bacteria count would be $1 \times (1.1)^5 = 1.61 > 1.50$.

So, we can conclude that during the interval of 4-5 minutes, the count would be more than 1.5 times the initial count.

The correct answer is Option C.

5.2.6 Ratio & Proportion

114. If we want to use 13.5 liters of red paint, the amount of yellow paint required $= \frac{5}{3} \times 13.5 = 22.5$ liters.

However, we only have 18 liters of yellow paint. So, we cannot use the entire 13.5 liters of red paint.

If we want to use 18 liters of yellow paint, the amount of red paint required $= \frac{3}{5} \times 18 = 10.8$ liters.

Since we have 13.5 liters of red paint, we can use the entire 18 liters of yellow paint.

So, total quantity of dye = 10.8 (red) +18 (yellow) = 28.8 liters.

The correct answer is Option C.

115. The ratio of the ingredients, i.e. Oil : Fragrance: Water = 1 : 8 : 3.

Thus, fragrance constitutes $\frac{8}{1+8+3} = \frac{8}{12} = \frac{2}{3}^{\text{rd}}$ of the total volume of the perfume.

=> The volume of the perfume $= \frac{3}{2} \times$ (volume of fragrance present).

We know that the volume of fragrance used = 60 ml.

Thus, total volume of the perfume $= \frac{3}{2} \times 60 = 90$ ml.

The correct answer is Option D.

116. We know that the lengths of three routes R, S, and P are in the ratio 3 : 4 : 3.

Let the lengths of the routes R, S, P be $3k, 4k$, and $3k$, respectively, where k is a constant of proportionality.

We know that the amount of fuel spent by the SUV, sedan and bike is in the ratio 6 : 2 : 3.

Let the amount of fuel spent by the SUV, sedan and bike be $6m, 2m$, and $3m$ respectively, where m is a constant of proportionality.

Thus, the ratio of the fuel efficiency, i.e. fuel consumed per unit distance traveled by the SUV, sedan and bike $= \frac{6m}{3k} : \frac{2m}{4k} : \frac{3m}{3k} = \frac{2m}{k} : \frac{m}{2k} : \frac{m}{k}$.

The constants $\frac{m}{k}$ can be cancelled from the ratio resulting in the required ratio to be $2 : \frac{1}{2} : 1$.

To convert the ratio to integer values, we multiply the terms of the ratio with 2 making the ratio $2 \times 2 : \frac{1}{2} \times 2 : 1 \times 2 = 4 : 1 : 2$.

The correct answer is Option A.

Alternate approach:

Since the ratio of the lengths of R, S, P are 3 : 4 : 3, we can, without loss of generality assume the lengths to be 3, 4, and 3, miles respectively.

Also, since the amount of fuel spent is in the ratio 6 : 2 : 3, we can, without loss of generality assume the amounts to be 6, 2, and 3 liters, respectively.

Thus, the ratio of the fuel efficiency i.e. fuel consumed per unit distance by the SUV, sedan and bike $= \frac{6}{3} : \frac{2}{4} : \frac{3}{3} = 2 : \frac{1}{2} : 1 = 4 : 1 : 2$.

117. Let x liters of the mixture be replaced with water.

Since the mixture has 80% milk, percentage of water = 100 – 80 = 20%.

Thus, quantity of water in the mixture = 20% of 40 = 8 liters.

The removed mixture of x liters also has 80% milk and 20% water.

Thus, quantity of water in x liters = 20% of $x = 0.2x$ liters.

Thus, quantity of water left when x liters of the mixture was removed = $(8 - 0.2x)$ liters.

Quantity of water added = x liters.

Thus, after addition, total volume of water = $(8 - 0.2x) + x = (8 + 0.8x)$ liters.

Also, total volume of the mixture remains at 40 liters (since we removed x liters and added back x liters).

The final volume of water now constitutes $(100 - 50) = 50\%$ of the total mixture volume.

Thus, we have $8 + 0.8x = 50\%$ of $40 \Rightarrow 8 + 0.8x = 20 \Rightarrow 0.8x = 12 \Rightarrow x = \frac{12}{0.8} = 15$.

Thus, the volume of mixture replaced = 15 liters.

The correct answer is Option B.

118. Let x liters of water be added to the existing mixture.

Since the mixture has 80% milk, percentage of water = 100 – 80 = 20%.

Thus, quantity of water in the mixture initially = 20% of 40 = 8 liters.

Quantity of water added = x liters.

Thus, after addition, total volume of water = $(8 + x)$ liters.

Also, total volume of the mixture = $(40 + x)$ liters.

The final volume of water now constitutes $(100 - 50) = 50\%$ of the total mixture volume.

Thus, we have

$8 + x = 50\%$ of $(40 + x) \Rightarrow 8 + x = 20 + 0.5x \Rightarrow 0.5x = 12$

$\Rightarrow x = \frac{12}{0.5} = 24$.

Thus, the volume of water added = 24 liters.

The correct answer is Option B.

Alternate approach:

Since the mixture has 80% milk, quantity of milk = 80% of 40 = 32 liters.

Since water was added to the mixture, the quantity of milk remained the same.

Thus, finally, this volume of milk constitutes 50% of the total mixture volume.

Hence, we can say that: 50% of (final volume of the mixture) = 32

=> Final volume of the mixture = $32 \times \frac{100}{50}$ = 64 liters.

The volume has increased since we added water to the mixture.

Thus, the volume of water added = 64 – 40 = 24 liters.

119. Let us assume that the total volume of the special container be 100 ml (since this is a percentage based question, 100 is convenient number; assuming any initial value will have no effect on the answer).

Thus, initial quantity of the chemical = 80% of 100 = 80 ml.

When 40% of the mixture is removed, remaining 100 – 40 = 60% of the mixture is left.

Thus, quantity of the chemical left = 60% of 80 = 48 ml.

Since only water is added to the remaining mixture, thereby making the mixture 100 ml., the quantity of chemical does not change.

Thus, finally, there is 48 ml chemical in 100 ml of the mixture.

Hence, percentage of chemical present = $\frac{48}{100} \times 100$ = 48%.

The correct answer is Option D.

5.2.7 Permutation and Combination

120. For the 3-digit number to be even, its unit digit must be either 2 or 8.

Number of ways, the unit place can be filled = 2 (either 2 or 8);

Number of ways, the tens place can be filled = 4; given that the repetition is allowed.

Number of ways, the hundreds place can be filled = 4

So, total number of ways $= 2 \times 4 \times 4 = 32$ ways.

The correct answer is Option C.

121. Since the letters and the digits can appear in any order, we must treat them as separate entities; thus we have to make a selection of out of 26 letters + 10 digits = 36 symbols.

Number of ways, each symbol can be chosen = 36;

So, total number of ways of choosing 3 symbols from 36 symbols for three places = 36^3; given that repetition is allowed.

The correct answer is option D.

122. Two 3s, appearing together, can be placed in the following ways in 5-digit numbers: 33XXX, X33XX, XX33X, and XXX33. Since '3' cannot appear again, we have 9 choices for each vacant place out of digits 0-9 (excluding the digit '3').

But wait! A 5-digit number cannot begin with 0; so for the 5-digit numbers that do not begin with '3' will have only 8 number of ways to its fill ten-thousandth place.

For numbers: X33XX, XX33X, and XXX33, total number of ways $= 3 \times 8 \times 9^2 = 24 \times 9^2$

For number: 33XXX, total number of ways $= 9^3$

Hence, total number of ways $= 9^3 + 24 \times 9^2$

The correct answer is Option D.

123. A triangle can be drawn taking any 3 vertices of a polygon.

So, the number of triangles $= C_3^7 = \dfrac{7 \times 6 \times 5}{1 \times 2 \times 3} = 35$

The correct answer is Option C.

124. Since one specific individual must be in the team, the question becomes:

"How many 2-member teams can be formed out of 5 individuals?"

Number of ways of choosing 2 individuals from 5 individuals $= C_2^5 = \dfrac{5 \times 4}{1 \times 2} = 10$.

The correct answer is Option B.

125. Number of ways of choosing only boys out of 8 boys = $C_2^8 = \frac{8 \times 7}{1 \times 2} = 28$

Number of ways of choosing only girls out of 10 girls = $C_2^{10} = \frac{10 \times 9}{1 \times 2} = 45$

Hence, the total number of ways = $28 + 45 = 73$;

Since this is a case of 'OR,' we would add the number of ways.

The correct answer is Option B.

126. We know that if there are total n objects, and out of these, there are p number of one type of identical objects, q number of another type of identical objects, and r number of another type of identical objects, then the number of ways these n objects can be arranged in a row to have unique patterns = $\frac{n!}{p! \times q! \times r!}$

There are a total 6 letters in BANANA; so $n = 6$, and there are 3 As and 2 Ns, so $p = 2$, and $q = 2$.

Thus, the answer is $\frac{6!}{3! \times 2!} = 60$.

The correct answer is Option B.

127. The number of ways of choosing 2 girls out of 12 girls = $C_2^{12} = \frac{12 \times 11}{1 \times 2} = 66$.

Similarly, the number of ways of choosing 2 boys out of 10 boys = $C_2^{10} = \frac{10 \times 9}{1 \times 2} = 45$.

Since this is a case of 'OR', we must add the number of ways, thus the answer is $66 + 45 = 111$.

The correct answer is 111.

128. We know that if there are total n objects, out of these there are p number of one type of identical objects, q number of another type of identical objects, and r number of another type of identical objects, then the number of ways these objects can be arranged in a row to have unique patterns = $\frac{n!}{p! \times q! \times r!}$

There are a total 6 letters in BANANA; there are 3 As and 2 Ns; since two 'Ns' have to be together, let's consider them as one letter; thus, we have the letters: B, (NN), A, A, A. So $n = 5$ and $p = 3$.

Thus, the answer is $\frac{5!}{3!} = 20$.

The correct answer is 20.

129. There are a total of 5 symbols.

$\square$ $\trianglerighteq$ $\otimes$ $\oslash$ $\Diamond$

We have take all 5 symbols together to form codes, considering that the $\square$ must NOT be at the beginning and the $\Diamond$ must NOT be at the end of the codes.

Number of ways 5 symbols can be arranged = $P_5^5 = 5! = 120$.

We will deduct the number of codes that start with a $\square$ and ends with a $\Diamond$; thus this asks for the arrangement of 3 symbols.

Number of ways 3 symbols can be arranged = $P_3^3 = 3! = 6$.

Thus, the desired number of codes = $120 - 6 = 114$.

The correct answer is 114.

5.2.8 Probability

130. Probability of an event = $\frac{\text{Number of favorable events}}{\text{Number of total events}}$

To get the required number of ways, we must NOT get both balls as green; we would deduct the number of ways we get both balls as green, upon drawing two balls from the total number of ways we can get two balls of any color.

Here, number of unfavorable cases = number of ways 2 green balls can be selected from 10 balls = C_2^{10}

Number of total cases = number of ways 2 balls can be selected from 30 balls = C_2^{30}

Hence, probability of NOT getting both balls as green

$$= 1 - \frac{C_2^{10}}{C_2^{30}} = 1 - \frac{(10 \times 9)/(1 \times 2)}{(30 \times 29)/(1 \times 2)} = 1 - \frac{3}{29} = \frac{26}{29}$$

The correct answer is Option D.

131. Since the coin is fair, it means that probabilities of getting a 'Head' or getting a 'Tail' upon a single toss are equal = 1/2.

Traditional approach of solving this question would be to find the probability for each time the coin turns 'Tail' and then add them, but that is a lengthy process.

An efficient approach would be to find out the probability of not getting a 'Tail' EVEN once, and then deducting it from total probability (1); this will assure that we get the 'Tails'—one time, two times, three, and all four times.

So, probability of getting at least one Head

= 1 – (Probability of getting 'Heads' in all 4 tosses)

$$= 1 - \left(\frac{1}{2} \times \frac{1}{2} \times \frac{1}{2} \times \frac{1}{2}\right) = 1 - \frac{1}{16} = \frac{15}{16}.$$

The correct answer is Option D.

132. f we solve this question through the traditional approach, it would be time-consuming as the problem can be considered solved if all three solved it; any two of them solved it; any one of them solved it; and then we will have to add the probabilities.

An efficient approach would be to calculate the probability of not solving the problem (it can be done in one way: only when none of them can solve it), and then deduct the probability from the total probability (1).

So, probability of not solving the problem

= P(Jack could not solve it) × P(Parker could not solve it) × P(Herman could not solve it)

$=> \left(1-\frac{4}{5}\right) \times \left(1-\frac{2}{3}\right) \times \left(1-\frac{3}{4}\right) = \frac{1}{5} \times \frac{1}{3} \times \frac{1}{4} = \frac{1}{60}$.

=> Thus, the probability of solving the problem $= 1 - \frac{1}{60} = \frac{59}{60}$.

The correct answer is Option C.

133. Total Number of actors = 5

Total number fo ways any three actors can be chosen =
$= C_3^5 = C_2^5 = 10$.

Let's calculate the number of ways Jack and Steve can be chosen together.

Since Jack and Steve are chosen, only one actor is left to be chosen, out of the remaining 3. One actor can be chosen out of 3 = 3 ways

Thus, total number of favorable selections = 10 − 3 = 7.

Thus, the desired probability $= \frac{7}{10}$

The correct answer is Option B.

134. Number of silver coins = p; Number of gold coins = q; so the total umber of coins = $p + q$;

The probability of getting 1st and 2nd coins silver = P (first silver coin) × P (second silver coin)

P (first silver coin) $= \frac{C_1^p}{C_1^{p+q}} = \frac{p}{p+q}$

After 1st draw, there would be only $(p+q-1)$ number of coins left.

So, P (second silver coin) $= \frac{C_1^{p-1}}{C_1^{p+q-1}} = \frac{p-1}{p+q-1}$

Hence, the probability of getting both coins as silver coins $= \left(\frac{p}{p+q}\right) \times \left(\frac{p-1}{p+q-1}\right)$

$= \frac{p(p-1)}{(p+q)(p+q-1)}$

The correct answer is Option B.

You could have solved this question by picking some values for p and q. Say $p = 2$ and $q = 4$; after solving, plug in the values of p and q in the options, and cross check which option matches the derived value. The derived value is 1/15 and matches with Option B.

135. Between 100 and 400, there are 400 − 100 + 1 = 301 numbers. The first few numbers divisible by '3' would be 102, 105, 108, ... and the few last ones would be .... 393, 396, 399. Similarly, the first few numbers divisible by '5' would be 100, 105, 110, ... and the few last ones would be .... 390, 395, 400. As you see that the number '105' is common, so we must take this into account.

The best way to exclude the duplicate numbers is to find the multiples of LCM of 3 & 5 = 15, thus the first few numbers divisible by '15' would be 105, 120, 135 .... and the few last ones would be .... 360, 375, 390.

Total numbers divisible by '3' between 100 & 400

$$= \left[\frac{\text{Last \# } - \text{ First \#}}{3} + 1\right] = \frac{(399 - 102)}{3} + 1 = 100 \text{ numbers.}$$

Total numbers divisible by '5' between 100 & 400

$$= \left[\frac{\text{Last \# } - \text{ First \#}}{5} + 1\right] = \frac{(400 - 100)}{5} + 1 = 61 \text{ numbers.}$$

Total numbers divisible by '15' between 100 & 400

$$= \left[\frac{\text{Last \# } - \text{ First \#}}{15} + 1\right] = \frac{(390 - 105)}{15} + 1 = 20 \text{ numbers.}$$

Thus, total numbers divisible either by '3' or by '5' = 100 + 61 − 20 = 141.

So, the probability of winning a gift = $\frac{141}{\text{Total number of coupons}}$ = 141/301

The correct answer is Option B.

136. Say the number of green balls is x, so the total number of balls = $(5 + x)$

Number of ways both the balls drawn will be green = C_2^x

Total number of ways 2 balls can be drawn = C_2^{5+x}

Hence, the probability of getting both the balls being green

$$= \frac{C_2^x}{C_2^{(5+x)}} = \frac{x(x-1)/1 \times 2}{(5+x) \times (4+x)/1 \times 2} = \frac{1}{6} \text{ (given)}$$

This reduces to $\frac{x(x-1)}{(5+x) \times (4+x)} = \frac{1}{6}$

We do not recommend you to solve the quadratic equation to get the value of x, you must plug in the values of x from the options and check if right hand side and left hand side are equal. If yes, the option is the correct answer.

Trying with $x = 4$ will satisfy the equation: $\frac{4 \times 3}{9 \times 8} = \frac{1}{6}$.

The correct answer is option B.

137. Understand the following table.

Sum	Ways	Number of ways	Freq. distribution & Probabilities
2	{1,1}	1	$p = \frac{1}{36}$
3	{2,1}, {1,2}	2	$p = \frac{1}{18}$
4	{3,1}, {2,2}, {1,3}	3	$p = \frac{1}{12}$
5	{4,1}, {3,2}, {2,3}, {1,4}	4	$p = \frac{1}{9}$
6	{5,1}, {4,2}, {3,3}, {2,4}, {1,5}	5	$p = \frac{5}{36}$
7	{6,1}, {5,2}, {4,3}, {3,4}, {2,5}, {1,6}	6	$p = \frac{1}{6}$
8	{6,2}, {5,3}, {4,4}, {3,5}, {2,6}	5	$p = \frac{5}{36}$
9	{6,3}, {5,4}, {4,5}, {3,6}	4	$p = \frac{1}{9}$
10	{6,4}, {5,5}, {4,6}	3	$p = \frac{1}{12}$
11	{6,5}, {5,6}	2	$p = \frac{1}{18}$
12	{6,6}	1	$p = \frac{1}{36}$
	Total number of ways	**36**	

Memorise...

1. **If Sum $\leq$ 7, then Number of ways = Sum − 1**
2. **If Sum $\geq$ 8, then Number of ways = 13 − Sum**

We have to calculate the probability of getting the sum > 10. It would be

$$\Rightarrow P(SUM > 10) = \frac{n\{\text{\# of ways of getting a sum of 11 or 12}\}}{36} = \frac{2+1}{36} = \frac{1}{12}$$

The correct answer is $\frac{1}{12}$.

138. We know that the probability of getting a 'Head' or a 'Tail' on any toss for a coin = 1/2.

We want Tail, Tail, and Head (in this order) on three tosses, so the probability of getting TTH = $(1/2) \times (1/2) \times (1/2) = 1/8$.

The correct answer is 1/8.

This question is different from a question asking for getting two Tails and a Head. In this case, one can get two Tails in I & II, or in I & III, or in II & III positions.
The answer would have been $3 \times 1/8 = 3/8$.

139. Probability of hitting the black parts = $\frac{\text{Area of the black parts}}{\text{Area of the outermost circle}}$

Area of the outermost circle = $\pi r^2 = \pi \times 4^2 = 16\pi$

Area of the innermost black circle (one of the black parts) = $\pi r^2 = 1^2 \times \pi = \pi$

Area of the 1-cm band (another black part) = $3^2 \times \pi - 2^2 \times \pi = 5\pi$

Total area of black parts = $\pi + 5\pi = 6\pi$

Thus, the probability of hitting the black parts = $\frac{6\pi}{16\pi} = 3/8$

The correct answer is 3/8.

5.2.9 Statistics

Questions on Average (Arithmetic Mean)

140. We know that arithmetic mean of x and y is 80,

=> $(x + y)/2 = 80$

=> $x + y = 160$ – – – – – (1)

Also, we know that arithmetic mean of x and z is 200,

=> $(x + z)/2 = 200$

=> $x + z = 400$ – – – – – (2)

To get the value of $(z - y)/8$, we subtract equation (2) from the equation (1) and get,

=> $z - y = 400 - 160 = 240$;

–> $(z - y)/8 = 240/8 = 30$

The correct answer is option C.

141. Say the three numbers are a, 7, and b; a being the largest and b being the smallest.

=> $(a + 7)/2 = 8$ => $a + 7 = 16$ => $a = 9$

Similarly, $(a + 7 + b)/3 = 6$

=> $9 + 7 + b = 18$

=> $b = 2$

=> the numbers are: 9, 7, & 2

So, the mean of the two smallest numbers = $(7 + 2)/2 = 4.5$

The correct answer is Option C.

142. Given,

Average = 7.5

=> Sum of 6 numbers = 6 × average = 6 × 7.5 = 45

Say the number multiplied by 4 is x, so instead of x, the number becomes $4x$. Or the sum of the numbers increases by $3x$.

=> New sum = $45 + 3x$

Given that new average = 7.5 – 0.5 = 7

=> New sum = 6 × new average = 6 × 7 = 42

=> $45 + 3x = 42$

=> $x = -1$

The correct answer is Option B.

Note that the question stem stated that after multiplying a number by 4, the average decreased rather than anticipated (increased), it implies that the number must be negative, thus only options A or B could be true.

143. Traditional approach:

Sum of scores of 5 tests = $5 \times 38 = 190$;

Say the score on the 6th test is x, so, sum of scores of 6 tests = $190 + x$;

=> Average score = $\frac{(190+x)}{6}$;

=> $41 = \frac{(190 + x)}{6}$; given that the average is 41.

=> $x = 56$

The correct answer is Option D.

Alternate approach:

You find that to increase the average score of 38 to 41, you need to increase it by 3. Since '3' is an average increase per test, so to get the total increase, you must multiply it by the total number of tests, i.e., 6.

=> Total increase = $3 \times 6 = 18$;

=> Desired score in the 6th test = $38 + 18 = 56$

144. Traditional approach:

Total rainfall in first x days = $32x$;

Total rainfall in $(x + 1)$ days = $(32x + 56)$;

=> Average rainfall in $(x + 1)$ days = $\frac{(32x + 56)}{(x + 1)}$

=> $36 = \frac{(32x + 56)}{(x + 1)}$; given that 'average rainfall in $(x + 1)$ days' = 36 mm

=> $36(x + 1) = (32x + 56)$

=> $36x + 36 = 32x + 56$

=> $4x = 20$

=> $x = 5$ days

The correct answer is option C.

Alternate approach:

Given that the increase in average = 36 − 32 = 4;

And difference of average of x days, 32 and today's rainfall, 56 = 56 − 32 = 24;

=> $24 = 4 \times (x + 1)$

=> $x = 5$ days

145. If all the elements of a set are multiplied by a constant, the mean also gets multiplied by the same constant.

However the mean remaining unchanged can happen only and only if mean = 0, so the Statement I is true.

Since sum of all the elements = Mean × # of elements

Sum of all the elements = 0, as mean = 0, so the Statement II is true.

Statement 3 is also not necessary because the set may contain all zeros.

The correct answer is Option C.

Questions on Median

146. First, arrange the Set X in an ascending order, we get: $-8, -2, 4, 6, \& \ 23$

=> Median = 4 (Middle-most term)

=> Median of Set Y = 4

Arranging Set Y in ascending order, we get: $-8, -2, 0, 8, 12, \& \ 13$; excluding y as we do not know its value. We know that Set Y has an odd number of terms, so there would be a unique term, equaling median value (here 4). Since there is no term, equaling 4, hence only y must be 4.

The correct answer is Option D.

147. Since there are odd number (7) of students, the median student would be $(7+1)/2 = 4^{th}$ student, and the corresponding height of that student would be the median height of seven students, when the bars are arranged in an ascending order, taking height into consideration.

[It is not wise to read the value of each column and jot down on the scratch pad and then after tagging them 1, 2, 3, ..., 7 in ascending order, pick up the 4th ranked value a median; it would be time-consuming.]

Arranging the students as per their heights: $4'3'', 4'6'', 4'9'', \mathbf{4'9''}, 5', 5'9'', 5'9''$. Since $\mathbf{4'9''}$ is the middle value, the student representing $\mathbf{4'9''}$ is the answer: Kim & Mike. Since Mike is not among options, Kim is the answer.

The correct answer is Option C.

148. We know that the Average (Mean) = $\frac{\text{Sum of heights}}{\text{Number of students}}$

$$\text{Average (Mean)} = \frac{4'3'' + 4'6'' + 4'9'' + 4'9'' + 5' + 5'9'' + 5'9''}{7}$$

$$= \frac{31'45''}{7} = \frac{(31 \times 12 + 45)''}{7} = \frac{417''}{7} = \frac{(417 \times 2.54)}{7} \text{ centimeters}$$

= 151.31 centimeters = ≈ 151 centimeters.

The correct answer is 151.

149. Range = Highest Value − Smallest Value

Thus, Range of heights = $5'9'' - 4'3'' = 1'6'' = 1 \times 12 + 6 = 18''$.

The correct answer is 18.

Note that 18/12 = 1.5 feet is a wrong answer. The answer needed is in inches.

150. Approach 1:

Median of all consecutive integers being a non-integer implies that the number of consecutive integers, X is an even number.

The median, 50.5 is derived by taking the average of 50 & 51.

The list looks like N1, N2, N3, ..., 50, 51, ...;

Since X is even, X/2 integers will be on the left of the median, and X/2 integers will be on the right of the median, 50.5. Counting of the smaller integer must start from 51 and we must deduct X/2 to reach the smallest, thus the smallest integer = 51 – X/2.

The correct answer is Option A.

Approach 2:

Since we already concluded that X is an even number, we can assume X.

Say X = 4. So, the list would be:

49, 50, 51, 52; smallest being 49

Only Option A, 51 – X/2 = 51 – 4/2 = 49 is qualified.

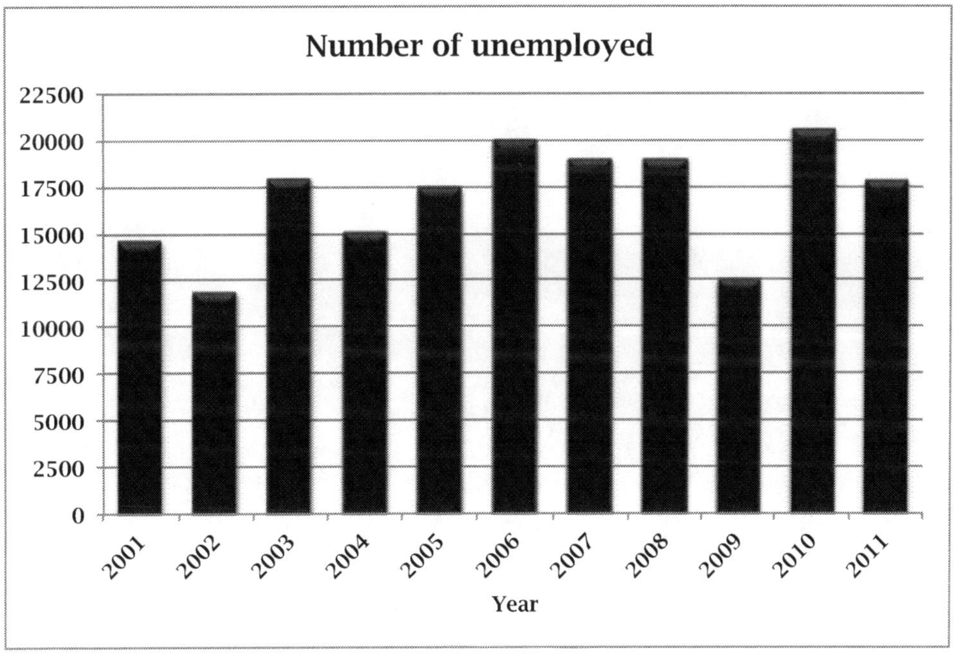

151. Since there are odd number (11) of years, the median year would be (11 + 1)/2 = 5th year, and the corresponding number of unemployed would be the median number of unemployed, when the columns are arranged in ascending order, taking the number of unemployed into consideration.

It is not wise to read the value of each column and jot down on the scratch pad and after tagging them 1, 2, 3, ..., 11 in ascending order, pick the 5th ranked value a median; it would be time-consuming.

Better approach would be to visually pick the 5th ranked bar as per the height of the bars, and then read the corresponding value—median.

Looking at the chart, we find that there are as many as 6 bars whose values are more than 17,500 (Y-axis). Among them, the smallest in height (5th ranked column) would be the answer. However we find that it is difficult to pin-point which of the two columns: year 2003 or year 2011 is the smallest. Well there is no need to be very particular about it as the questions asks for the closest value of median, and does not ask for the median year.

Read the data of year 2003 and year 2011.

The least count of Y-axis is 2,500; we can approximate the values of year 2003 and year 2011 to be 18,000. So the median number of unemployed equals 18,000.

The correct answer is option D.

Questions on Average and Median

152. Given that,

Median defect per batch = Average defect per batch; for nine batches;

Let us first deduce the median.

By arranging the eight values of defects in ascending order, we get;

2, 3, 3, 4, 5, 6, 7, 8

From the options, if the # of defects in the ninth batch is 2, 3, or 4, the median would be 4—Middle-most value.

However if it is either 5, 6, 7 or 8, the median would be 5—Middle-most value, so we are not sure about the value of median, but it would be either 4 or 5.

Let us calculate the average now. It must be either 4 or 5 as it is given that average = median.

Say the # of defects in the ninth batch is x,

=> Average = $(2 + 3 + 3 + 4 + 5 + 6 + 7 + 8 + x)/9 = (38 + x)/9$;

=> $(36 + 2 + x)/9$

=> $4 + \frac{(2+x)}{9}$

$\frac{(2+x)}{9}$ must be 0 or 1; thus x must be -2 or 7.

Since # of defects cannot be negative, $x = 7$.

The correct answer is option D.

153. Mode of any data series is the value of the characteristic that occurs most often among all the values. Here, the maximum number of students for a particular score is 12, thus the modal score representing those 12 students = 5.

The correct answer is Option B.

154. We know that the Average (Mean) = $\frac{\text{Sum of heights}}{\text{Number of students}}$

=> $\frac{\text{Score '0'} \times \text{\# of students scoring '0'} + \text{Score '1'} \times \text{\# of students scoring '1'} + ... + \text{Score '10'} \times \text{\# of students scoring '10'}}{71}$

=> $\frac{(0 \times 3 + 1 \times 4 + 2 \times 6 + 3 \times 7 + 4 \times 8 + 5 \times 12 + 6 \times 11 + 7 \times 9 + 8 \times 6 + 9 \times 3 + 10 \times 2)}{71}$

$= \frac{353}{71} = 4.97.$

The correct answer is 4.97.

Question on Range

155. We do not know the values of $a, b,$ & c.

So, Minimum value of range = 76 – (−43) = 119; if any of a, b, & c is greater than 76 and/or any of $a, b,$ & c is smaller than − 43, the range would be greater than 119.

Since range is always positive, so Statement I is ruled out.

Statement II is also ruled out as its value (112) is less than 119; we know that the minimum value of range should be 119.

The correct answer is Option C.

Questions on Standard deviation

156. Adding or subtracting a constant number from all elements of a set does not affect the standard deviation (SD) of the set. We see that all sets P, Q and R can be obtained by adding or subtracting a constant from the elements of the given Set X; hence all the sets have equal standard deviations (SD).

The correct answer is Option D.

Note that, if you multiply each element of a set by a constant 'm', the SD would be m times the original SD. Say, each element of the set is multiplied by 3, then the SD would be 3 times the original SD = 3SD.

Similarly, if you divide each element of the set by a constant 'm', the SD would be $1/m$ times the original SD = SD/3.

Multiplying with or dividing by a negative number does not affect SD's sign. SD is always a positive number.

157. Adding or subtracting a constant from each element in the set has no effect on standard deviation; however multiplication and division does change SD.

Approach 1:

Looking at options A, B, & C, we find that a constant term is NOT added to or NOT subtracted from each element, this implies that SD of any of the set would not be the same as the SD of Set X. So only option left is: D—correct answer.

Though at first sight, it does not look that Option D satisfies the criterion, expanding each element will do the trick.

Option D. $\{p + qr, q(1 + r), r(q + 1)\}$

- I element: $p + qr$

- II element: $q(1 + r) = q + qr$
- III element: $r(q + 1) = r + qr$

You must have observed that each element is derived by adding a constant qr to each element of set X. This implies that SD of the set would be the same as that of set X.

The correct answer is Option D.

Approach 2:

Say, $p = 1, q = 2$ & $r = 3$. No need to calculate SD.

Let us analyze the options by plugging in the values of p, q & r. The values would be as follows.

(A) 2, 2, 3

(B) 1, 3, 3

(C) 1, 2, 4

(D) **7, 8, 9**: Constant '6' is added to $\{p = 1, q = 2, r = 3\}$, implying same SD.

158. Since SD is always positive, so statements I & II are ruled out.

SD for the given set CANNOT be '0' as there are deviations among elements. For SD to be '0', all elements of a set must be equal.

The correct answer is Option D.

159. We know that average number of stickers = $\frac{\text{total number of stickers}}{\text{number of friends}}$

Thus, total number of stickers = (number of friends)$\times$(average number of stickers)

Since the average number of stickers with the five friends was 15, the total number of stickers with them = $5 \times 15 = 75$.

The maximum number of stickers with any friend = 18.

In order that one friend should have the minimum number of stickers, we need to maximize the number of stickers with the remaining three friends (since one of them already has 18 stickers).

Thus, we assume that the number of stickers with three of the friends is 17, 16 and 15 (since the number of stickers with any friend needs to be different, we cannot assume that the three friends have 17 stickers each).

Thus, the number of stickers with the last friend = $75 - (18 + 17 + 16 + 15) = 75 - 66 = 9$.

Hence, the minimum number of stickers with any friend = 9.

The correct answer is Option C.

5.2.10 Data Analysis

160. From the bar chart, we know that the total government spending is $200 + 500 + 100 + 200 = \$1,000$ million dollars.

From the pie chart, total government spending was 20% of GDP.

So, the GDP was $\frac{1,000}{20\%} = 5,000$ million dollars. Since the government's cost of operation was 200 million, it was $\frac{200}{5,000} \times 100\% = 4\%$ of the total GDP.

The correct answer is Option A.

161. From the question above, the GDP is $5,000$ million dollars, so the dollar value of Management Consulting Services (= Miscellaneous items) is 3% of $5,000 = 3 \times \frac{5,000}{100} = 150$ Million dollars, and Services account for the 25% or one-forth of the total GDP, or $\frac{5,000}{4} = 1,250$ Million dollars. So, the fraction of Management Consulting Services in the Services is $\frac{150}{1,250} = 0.12$, or 12%.

The correct answer is Option C.

Alternatively,

We see that Special Management Consulting (= Miscellaneous items) would be 3% of GDP; however, Services is 25% of GDP. So, required percent = $\frac{3}{25} \times 100 = 12\%$

162. The total advertising cost for 4 weeks = 10% of (Average length of a data request per week × Total number of data request for 4 weeks)
$= 10\% \times 4 \times (20,000 + 20,000 + 15,000 + 5,000) = 10\% \times 4 \times 60,000 = 24,000$ dollars.

The correct answer is Option D.

163. From the chart we know that Michelle's spending on travel is 19%, thus Michelle's total annual dollar spending = $\frac{100}{19} \times 38,000 = 200,000$. We know that Michelle's spending on clothing is 15%, thus her spending on clothes was 15% of $200,000 = \$30,000$.

The correct answer is Option D.

164. The average sale will be calculated by dividing the total quarter-wise sale by the # of products (10), and then we can decide which quarter is the correct answer.

$$\text{Average} = \frac{\text{Total sale in a quarter}}{10}$$

Since the denominator for each quarter is a constant (10), only the numerator will determine the quarter with the highest average; the quarter, which has the highest 'total sale' would be the answer. Here, Quarter I total sale = 175 MT (see the last row of the table) is highest among all quarters.

The correct answer is Option A.

165. We have to select only one statement that is correct.

Let us analyze each statement one by one.

(A) **The statement:** In Quarter I, there is only one product that has the highest sale.

This is **incorrect** because in Quarter I, there are as many as 3 products, and not only one, that have the highest sale (24 MT): CR pipe, HR pipe, & HR wire.

(B) **The statement:** In Quarter II, for 'HS billet' and 'MS billet', the ratio of 'sale to inventory' is 0.

This is **incorrect** because in Quarter II, for 'HS billet' and 'MS billet', the ratio of 'sale to inventory' is NOT 0; it is indeterminable or undefined. (We cannot determine the value of a number divided by 0)

In fact, for 'HS billet' and 'MS billet', the ratio of 'inventory to sale' (inverse) is 0, so read attentively.

(C) **The statement:** In Quarter III, the product that has the highest inventory also has the least sale.

This is **incorrect** because in Quarter III, the product 'HR pipe' has the highest inventory (21 MT), but the product 'HR sheet' has the least sale (0).

(D) **The statement:** In Quarter IV, the product that has the highest inventory also has the greatest sale.

This is **correct** because in Quarter IV, the product 'CR sheet' has the highest inventory (21 MT) as well as the greatest sale (15 MT); though the greatest sale is shared by two more products, unlike Option A statement, Option D statement does imply that 'CR sheet' is the only product to have the greatest sale.

The correct answer is Option D.

166. We want to keep the per unit cost of the item least. The curve shows that the least per unit cost is approximately \$0.80 (bottom of the curve). The corresponding output for the same would be any thing between 2,000 to 2,100.

The correct answer is Option C.

167. The per unit cost at 1,200 units of output is approximately \$1.50. If we extend the horizontal line from 1.50, we meet the curve at a point which has the corresponding output equal to approximately 3,000.

The correct answer is option C.

168. First, locate \$125M in the X-axis, and then extent a line vertically to meet the trend-line; from that intersecting point, read the value at the Y-axis, which is approximately \$28M.

The correct answer is Option B.

169. Let us assume that the linear equation is $p = mt + c$, where m & t are constants.

Since there are two unkowns m & c, we must have two values of two corresponding points from the graph to get the values of m & c.

Let us choose I point @ \$70M (t); the corresponding profit = \$5M (p); thus,

=> $5 = 70m + c$.....(1);

Let us choose II point @ \$130M (t); the corresponding profit = \$30M (p); we have chosen this point as the values are easier to read for \$130M. thus,

=> $30 = 130m + c$.....(2);

By subtracting equation (2) from (1), we get,

=> $60m = 25$ => $m = 25/60 = 0.42$.

By plugging in the value of $m = 0.42$ in equation (1), we get $c = -24.4$.

Thus, the equation is $p = 0.42t - 24.4$.

The correct answer is Option B.

Alternate approach:

Let us pick any point on the line, say $p = 5$ and $t = 70$.

We substitute the above point in each equation and verify which holds true.

Only in Option B: $p = 0.42t - 24.4$, we have

RHS = $0.42 \times 70 - 24.4 = 5$ = LHS.

170. Total number of complaints in the month of January

= # of complaints in plant 1 + # of complaints in plant 2 + # of complaints in plant 3

= 20 + 22 + 12 = 54.

Similarly, total number of complaints in the month of June

= # of complaints in plant 1 + # of complaints in plant 2 + # of complaints in plant 3

= 10 + 12 + 22 = 44.

Thus, the total number of complaints in three plants in the month of January is more than that in the month of June by $54 - 44 = 10$.

The correct answer is Option A.

171. In the graph, the symbol used for Plant 3 is a triangle. The best approach to figure out the answer is by observing that for how many months the triangles are placed above the symbols used for Plants 1 (Rhombuses) and for Plant 2 (Squares).

We see that for the months—April, May, June, July, October, November, and December—the triangles are placed above the rhombuses and above the squares; thus there are 7 such months.

The correct answer is 7.

172. 'Entertainment' expenses for a family can be calculated as "(% of expenses) X (Total expenses by the family)".

The % expenses on 'Entertainment' head by the families are:

(A) Family A: $(100\% - 70\%) = 30\%$;

(B) Family B: $(100\% - 65\%) = 35\%$ (Highest among the families);

(C) Family C: $(100\% - 75\%) = 25\%$

The total expenses by the families are:

(A) Family A: $2,500$;

(B) Family B: $2,800$;

(C) Family C: $3,500$ (Highest among the families);

The total expenses on 'Entertainment' head by each family can be calculated as:

(A) Family A: $30\% \times 2,500 = \$750$;

(B) Family B: $35\% \times 2,800 = \$980$ (Highest among the families)—correct answer;

(C) Family C: $25\% \times 3,500 = \$875$

The correct answer is Option B.

173. 'Clothing' expenses for a family can be calculated as "% of expenses × Total expenses by the family".

The % expenses on 'Clothing' head by the families are:

(A) Family A: $(30\% - 15\%) = 15\%$ (Least among the families);

(B) Family B: $(50\% - 30\%) = 20\%$;

(C) Family C: $(50\% - 30\%) = 20\%$

The total expenses by the families are:

(A) Family A: $2,500$ (Least among the families);

(B) Family B: $2,800$;

(C) Family C: $3,500$

Since Family A is least for both the parameters: % expenses on 'Clothing' & total expenses, its expenses on clothing would be the correct answer.

Expenses on clothing by Family A: $15\% \times 2,500 = \$375$

The correct answer is 375.

174. The question asks us for which year, the

'Sum of # of commuters for Intra-City Bus + Local Train + Private taxi'

is the highest.

Looking at the graph and the options, it seems that we may have to deal with 12 data points (4 options × 3 bars per year), but it is not so; mere observation can get us the answer.

We see that the black color bar is highest for years 2005 & 2009.
Similarly, the dark grey color bar is highest for years 2007 & 2009.
Similarly, the light grey color bar is highest for year 2009.

Observing the graph, the Intra-City Bus had its peak in 2005 and 2009, the Local Train in 2007 and 2009, and the Private Taxi in 2009. Since 2009 appears consistently across all three, it is the year with the highest total commuters. Since the question wants us to find out the year and not the value, we must not waste our time in calculating it.

The correct answer is Option D.

175. The question asks us to calculate the

'Sum of # of commuters for Intra-City Bus + Local Train + Private taxi' for all the years and answer with the least value.

Like the question above, we can solve this one with mere observation.

We see that the black color bar is lowest for years 2006, 2007, & 2008.
Similarly, the dark grey color bar is lowest for year 2008.
Similarly, the light grey color bar is lowest for years 2005, 2006, & 2008.

We see that year 2008 is common in all three parameters; thus,

'Sum of # of commuters for Intra-City Bus + Local Train + Private taxi' for year 2008 must be the correct answer.

'Sum of # of commuters for Intra-City Bus + Local Train + Private taxi' for year 2008 = 325 + 275 + 150 = 750M.

The correct answer is Option C.

5.3 Advanced Math

5.3.1 Polynomials

176. Given that

$(2x^5 - 3x^4 - 4x^2 + 3) - (2x^4 - 3x^2 + 4x + 13) - (2x^3 - 3x^2 - 4x)$

After opening bracket, we get,

$2x^5 - 3x^4 - 4x^2 + 3 - 2x^4 + 3x^2 - 4x - 13 - 2x^3 + 3x^2 + 4x$

$2x^5 - x^4(3 + 2) - x^2(4 - 3 - 3) - x(4 - 4) + 3 - 13$

$2x^5 - 5x^4 + 2x^2 - 10$

The correct answer is Option B.

177. Given that

$13x^5y^{-3}z^2 \times 4x^{-2}y^3z^{-3}$

We know that when we multiply or divide algebraic terms, their exponents are added or subtracted.

Thus, $13x^5y^{-3}z^2 \times 4x^{-2}y^3z^{-3} = 52x^{5-2}y^{-3+3}z^{2-3}$

$52x^3y^{\circ}z^{-1} = 52x^3z^{-1}$; as we know that $a^{\circ} = 1$.

The correct answer is Option B.

178. Given that

$(-3x^4 + 3) \times (2x^4 + 4x)$

When we multiply two expressions, each term of an expression gets multiplies with each term of another expression.

Thus, $(-3 \times 2 \times x^{4+4}) + (3 \times 2x^4) - (3 \times 4 \times x^{4+1}) + (3 \times 4x)$

$= -6x^8 + 6x^4 - 12x^5 + 12x$

$= -6x^8 - 12x^5 + 6x^4 + 12x$

The correct answer is Option B.

179. Given that: $\frac{6x^5-33x^4+21x^3-6x^2-30x}{3x}$

=> $\frac{\cancel{3x}(2x^4 - 11x^3 + 7x^2 - 2x - 10)}{\cancel{3x}}$; taking $3x$ common.

=> $2x^4 - 11x^3 + 7x^2 - 2x - 10$

The correct answer is Option B.

180. Given: $\dfrac{3x^3 - 24x^2 - 4x + 32}{x - 8}$

=> $\dfrac{(3x^3 - 24x^2) - (4x - 32)}{x - 8}$; making pairs of terms

=> $\dfrac{3x^2(x - 8) - 4(x - 8)}{x - 8}$; anticipate that the denominator may have factor $(x - 8)$

=> $\dfrac{(3x^2 - 4)\cancel{(x - 8)}}{\cancel{(x - 8)}}$

=> $3x^2 - 4$

The correct answer is Option C.

181. Given: $\dfrac{8x^3 + 125}{2x + 5}$

=> $\dfrac{(2x)^3 + (5)^3}{2x + 5}$

We know that $a^3 + b^3 = (a + b)(a^2 - ab + b^2)$

Thus, $\dfrac{(2x)^3 + (5)^3}{2x + 5} = \dfrac{\cancel{(2x + 5)}(4x^2 - 10x + 25)}{\cancel{(2x + 5)}}$

$= 4x^2 - 10x + 25$

The correct answer is Option C.

182. Given: $\dfrac{8 - 729x^6}{\sqrt{2} - 3x}$

=> $\dfrac{(2)^3 - (9x^2)^3}{\sqrt{2} - 3x}$; we know that $a^3 - b^3 = (a - b)(a^2 + ab + b^2)$

$= \dfrac{(2 - 9x^2)(4 + 18x^2 + 81x^4)}{\sqrt{2} - 3x}$

=> $\dfrac{[(\sqrt{2})^2 - (3x)^2][4 + 18x^2 + 81x^4]}{\sqrt{2} - 3x}$; we know that $a^2 - b^2 = (a + b)(a - b)$

Thus by canceling the term $(\sqrt{2} - 3x)$, we get

$$\frac{[\cancel{(\sqrt{2} - 3x)}(\sqrt{2} + 3x)][4 + 18x^2 + 81x^4]}{\cancel{(\sqrt{2} - 3x)}}$$

=> $(\sqrt{2} + 3x)(4 + 18x^2 + 81x^4)$

The correct answer is Option D.

5.3.2 Quadratic equations

183. Simplifying the given equation, we have

$\frac{3}{x-1} - \frac{2}{2x-3} = 1$

$=> \frac{3(2x-3)-2(x-1)}{(x-1)(2x-3)} = 1$

$=> \frac{6x-9-2x+2}{2x^2-3x-2x+3} = 1$

$=> \frac{4x-7}{2x^2-5x+3} = 1$

$=> 4x - 7 = 2x^2 - 5x + 3$

$=> 2x^2 - 9x + 10 = 0$

$=> 2x^2 - 4x - 5x + 10 = 0$

$=> 2x(x - 2) - 5(x - 2) = 0$

$=> (2x - 5)(x - 2) = 0$

$=> x = \frac{5}{2}$ or 2.

Thus, the integer value of x that satisfies the given equation is 2.

The correct answer is 2.

184. $\frac{3}{x} + \frac{x}{6} = \frac{3}{2} => \frac{3\times6+x\times x}{6x} = \frac{3}{2} => \frac{(18+x^2)}{6x} = \frac{3}{2} => 18 + x^2 = \frac{3}{2} \times 6x$

$=> 18 + x^2 = 9x$

$=> x^2 - 9x + 18 = 0$

$=> x^2 - 6x - 3x + 18 = 0$

$=> x(x - 6) - 3(x - 6) = 0$

$=> (x - 3)(x - 6) = 0$

$=> x = 3$ or 6.

Thus, sum of the values of x that satisfy the given equation is 3 + 6 = 9.

The correct answer is Option D.

Alternate approach:

We know that for a quadratic equation $ax^2 + bx + c = 0$, the 'Sum of roots $= -\frac{b}{a}$', thus for the equation, $x^2 - 9x + 18 = 0$, $a = 1$, & $b = -9$.

=> Sum of roots $= -\frac{b}{a} = -(\frac{-9}{1}) = 9$.

185. Simplifying the equation, we have

$\frac{12}{x+4} + \frac{6}{x-6} = 4$

=> $\frac{12(x-6)+6(x+4)}{(x+4)(x-6)} = 4$

=> $\frac{12x-72+6x+24}{x^2+4x-6x-24} = 4$

=> $\frac{18x-48}{x^2-2x-24} = 4$

=> $\frac{2(9x-24)}{x^2-2x-24} = 4$

=> $\frac{9x-24}{x^2-2x-24} = 2$

=> $9x - 24 = 2(x^2 - 2x - 24)$

=> $9x - 24 = 2x^2 - 4x - 48$

=> $2x^2 - 13x - 24 = 0$

=> $2x^2 - 16x + 3x - 24 = 0$

=> $2x(x - 8) + 3(x - 8) = 0$

=> $(2x + 3)(x - 8) = 0$

=> $x = -\frac{3}{2}$ or 8.

Thus, the integer value of x is 8.

The above is a lengthy and time-consuming process; a better approach would be to plug-in option values in the given equation. We can see that if $x = 8$, the equation is satisfied and hence it is the solution.

The correct answer is Option C.

186. Let us solve the equations and find out the roots.

$x^2 - 3x + 2 = 0$

=> $x^2 - 2x - x + 2 = 0$

=> $x(x - 2) - 1(x - 2) = 0$

=> $(x - 1)(x - 2) = 0$

=> $x = 1$ or 2

$y^2 - 8y + 15 = 0$

=> $y^2 - 5y - 3y + 15 = 0$

=> $y(y - 5) - 3(y - 5) = 0$

=> $(y - 5)(y - 3) = 0$

=> $y = 5$ or 3

On comparing the roots, we see that $y > x$ for both values of x and y.

The correct answer is option B.

Graphic polynomial and quadratic equations

187. The given graph is $f(x) = (x - 1)^2 + 3$.

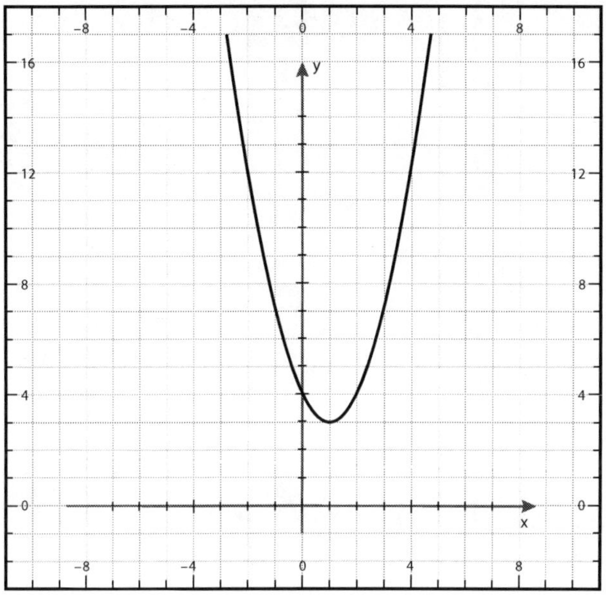

We need to find the graph of $g(x) = (x + 1)^2 + 1$.

Here, we see that in $f(x) = (x - 1)^2 + 3$, the term $(x - 1)$ has been replaced with $(x + 1)$, i.e., x has been replaced with $(x + 2)$.

Also, the constant 3 has been replaced by 1 i.e. the y values of the graph have been reduced by $3 - 1 = 2$ units.

Thus, the graph would shift by 2 units left (since x has been increased by 2) and would shift by 2 units down (since the constant has been reduced by 2).

To find the correct graph, let us pick one reference point on the original graph.

Let the point be the one where $f(x)$ has the lowermost point i.e. $(1, 3)$.

Thus, on shifting, this point would move to $[(1 - 2),\ (3 - 2)] => (-1, 1)$.

Thus, the graph of $g(x)$ should have the minimum position at $(-1, 1)$.

This is satisfied by the graph shown in Option D.

The correct answer is Option D.

188. For comparing the graphs of $f(x)$ and $g(x)$, we need to pick some reference point.

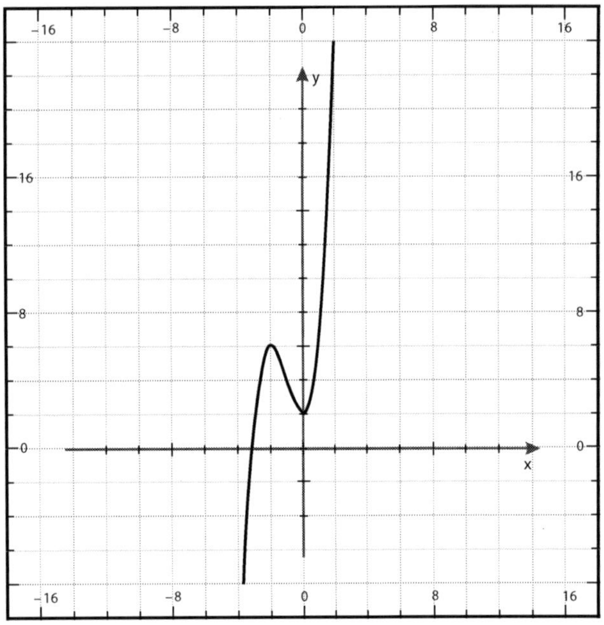

$y = f(x)$

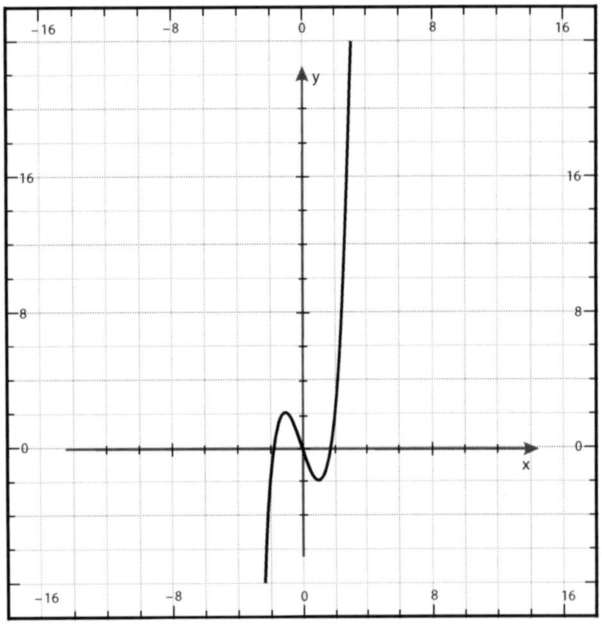

$y = g(x)$

The graph of $f(x)$ has a small peak on the left and a small valley on the right. We can pick either of the two points as our reference point.

Let us work with the peak on the left.

In the graph of $f(x)$, the point of the peak is $(-2, 6)$ (we should note that each gap along the X or Y axis is of 2 units).

The same point in the graph of $g(x)$ is $(-1, 2)$.

Thus, we can say that the point has shifted to the right by 1 unit and down by 4 units.

Thus, in order to shift the graph right by 1 unit, we should replace x with $(x - 1)$ and in order to shift the graph down by four units, we should subtract 4 overall.

Thus, Option D satisfies the above conditions.

The correct answer is Option D.

189. If we compare the graphs of $f(x)$ and $g(x)$, it seems that $g(x)$ is the inverted form of $f(x)$ i.e. reflected about the X-axis.

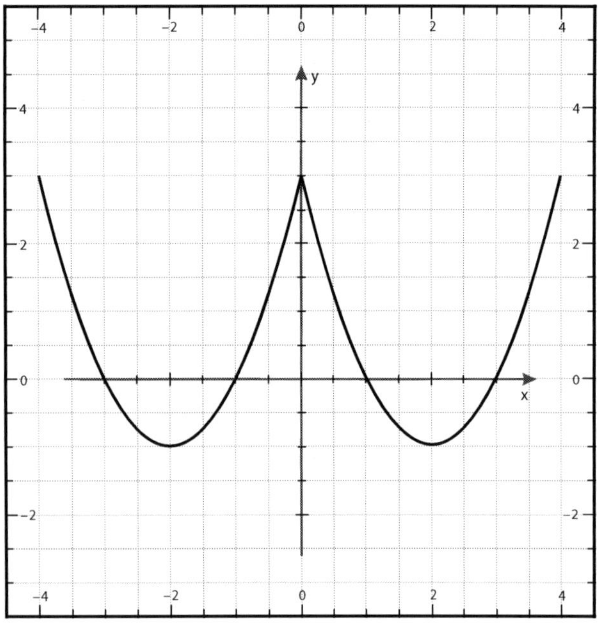

$y = f(x)$

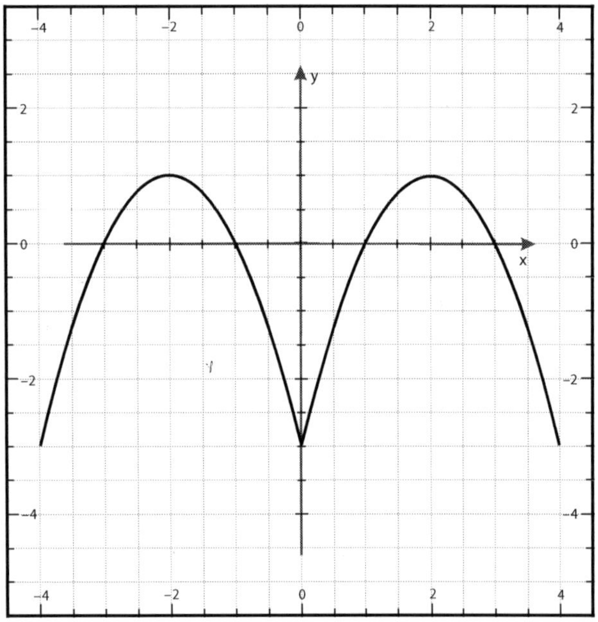

$y = g(x)$

To verify this, we need to pick some reference points.

The graph of $f(x)$ has a peak on the Y-axis and two valleys, one of the left and the other on the right.

The coordinates of the above points are $(0,3)$, $(-2,-1)$, and $(2,-1)$, respectively.

In the graph of $g(x)$, we see that peak has become the valley and the valleys have become peaks. The coordinates of the above points are $(0,-3)$, $(-2,1)$, and $(2,1)$, respectively.

Thus, we find that for each of these points, the value of x remained unchanged while the value of y has been negated.

Thus, we can say that $g(x)$ is the negation of $f(x)$ i.e. $g(x) = -f(x)$.

Thus, Option C satisfies the above condition.

The correct answer is Option C.

190. If we compare the graphs of $f(x)$ and $g(x)$, it seems that $g(x)$ has been obtained by reflecting $f(x)$ about the Y-axis.

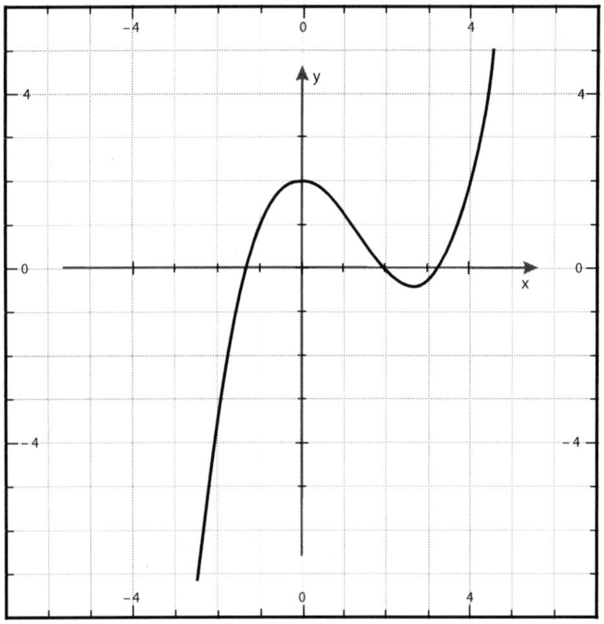

$y = f(x)$

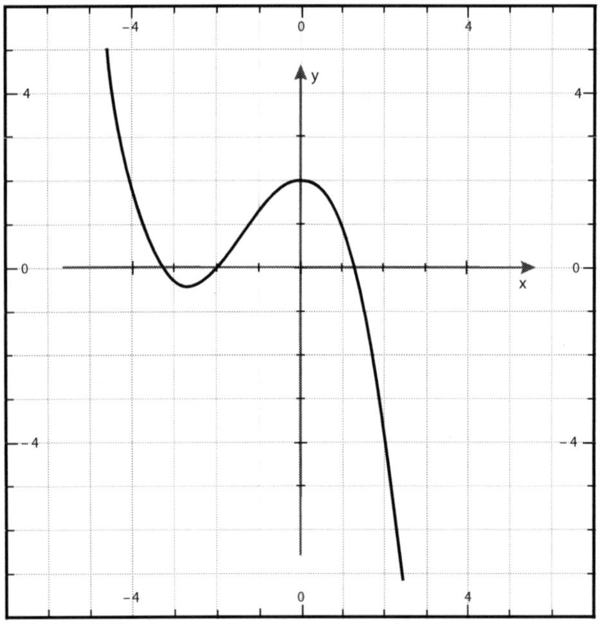

$y = g(x)$

To verify this, we need to pick some reference points.

The graph of $f(x)$ has a peak on the Y-axis and intersects the X-axis at two points on the right.

The coordinates of the peak and the first point of intersection with the X-axis on the right are $(0, 2)$ and $(2, 0)$, respectively.

In the graph of $g(x)$, we see that peak has remained the same and the point of intersection has now come to the left at $(-2, 0)$.

Thus, we find that the value of y remained unchanged while the value of x has been negated.

Thus, we can say that $g(x) = f(-x)$.

Thus, Option B satisfies the above conditions.

The correct answer is Option B.

191. If we compare the graphs of $f(x)$ and $g(x)$, it seems that $g(x)$ is obtained by reflecting $f(x)$ about the X-axis i.e. x in $f(x)$ has been replaced by $(-x)$.

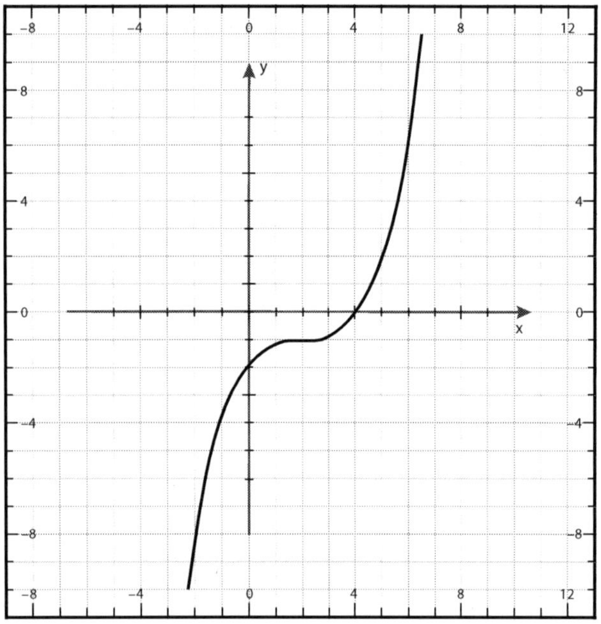

$y = f(x)$

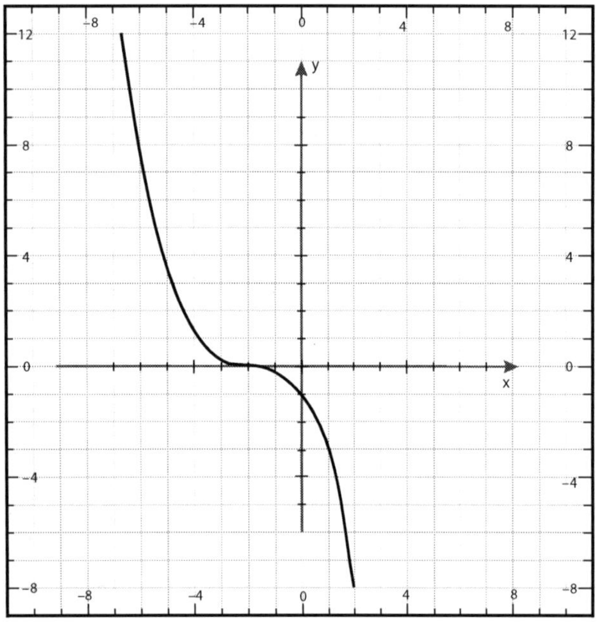

$y = g(x)$

Also, we can observe that after reflection, the graph has been shifted up.

To verify this, let us select a point on $f(x)$, where it intersects the Y-axis.

The coordinates of the above point is $(0, -2)$.

The coordinates of the point where $g(x)$ intersects the Y-axis is $(0, -1)$.

Thus, it is clear that the graph of $f(x)$ has been shifted up by 1 unit after reflecting about the Y-axis.

Hence, we can say that $g(x) = f(-x) + 1$.

Thus, Option B satisfies the above condition.

The correct answer is Option B.

192. $g(x) = \frac{x^2}{3} + 3 = \frac{1}{3}(x^2 + 3) + 2$

$=> g(x) = \frac{1}{3} \times f(x) + 2$

Thus, we see that $f(x)$ has been multiplied with $\frac{1}{3}$ and then 2 has been added.

Thus, the values of y corresponding to different x values in $\frac{1}{3} \times f(x)$ will be $\frac{1}{3}^{\text{rd}}$ of the corresponding values of y in $f(x)$.

Hence, the nature of the curve in will be more flat as compared to $f(x)$ and the y values corresponding to same x values would be lower than in $f(x)$.

For example, in $f(x)$, the point where it intersects the Y-axis, the coordinates of the point is $(0, 3)$.

The corresponding value of the coordinates where $\frac{1}{3} \times f(x)$ intersects the Y-axis are $(0, \frac{1}{3} \times 3) = (0, 1)$ as shown below.

Also, to explain the flatness of the curve, let us observe the value of y in the graph of $f(x)$ when $x = 3$ and the corresponding value of y in the graph of $\frac{1}{3} \times f(x)$ at the same value of x.

We can see that in $f(x)$, the value of y is 12 while in $\frac{1}{3} \times f(x)$, the value is 4, i.e. the value has become $\frac{1}{3}^{\text{rd}}$ of the initial value.

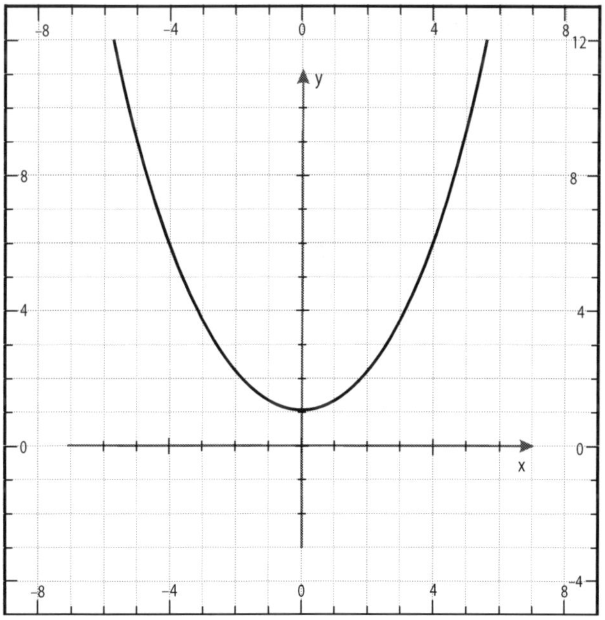

$y = \frac{1}{3} \times f(x)$

Now, since $g(x) = \frac{1}{3} \times f(x) + 2$, the above graph will shift up by 2 units as shown below.

Thus, the coordinates of the point of intersection of $f(x)$ and $\frac{1}{3} \times f(x) + 2$ will be the same i.e. $(0, 3)$.

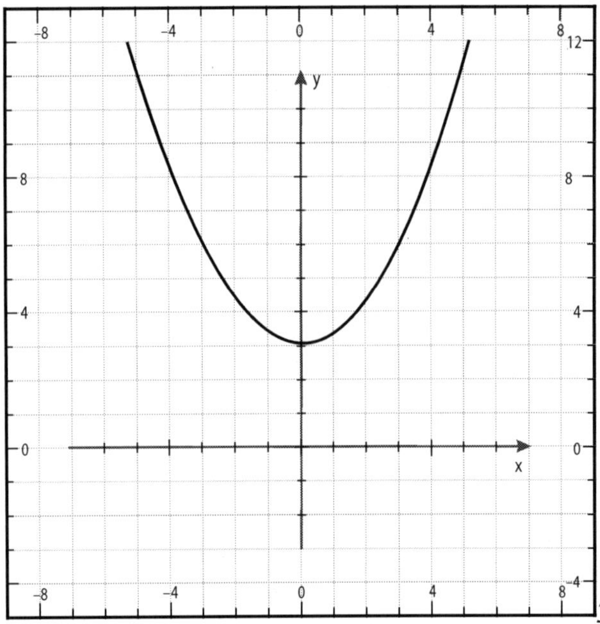

$y = \frac{1}{3} \times f(x) + 2$

Thus, the overall effect is that of making the graph more flat, i.e. wider without shifting the graph of $f(x)$.

The correct answer is Option D.

193.

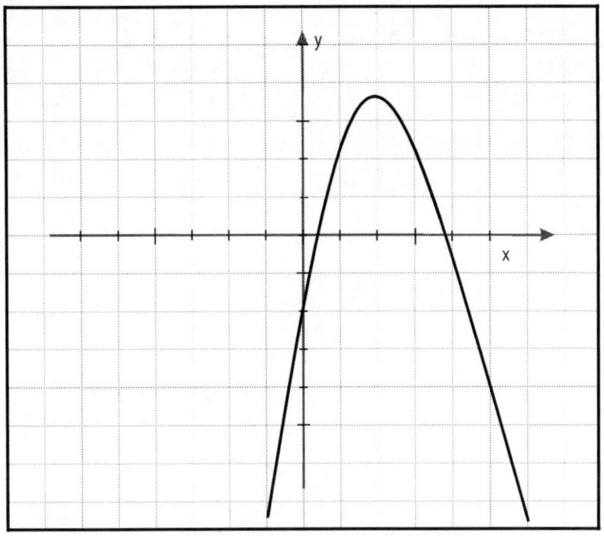

From the graph above, when $x = 0$, the value of y i.e. $f(0)$ is negative since the graph intersects the Y-axis below the origin.

Substituting $x = 0$ in $f(x)$, we get $f(0) = 0 + 0 + c = c$. Thus, the value of c must be negative.

Again, the graph intersects the X-axis at two points (the roots), both of which are on the positive X-axis. Thus, both the roots of the quadratic are positive. Hence, product of the roots is also positive.

We know that the product of the roots of the quadratic function:

$f(x) = ax^2 + bx + c$ is given by $\frac{c}{a}$.

Thus, we can say that $\frac{c}{a}$ is positive.

Since we know that c is negative, a must be negative as well.

Again, since both roots of the above quadratic are positive, the sum of the roots will also be positive.

We know that the sum of the roots of the quadratic function:

$f(x) = ax^2 + bx + c$ is given by $-\frac{b}{a}$.

Thus, we can say that $-\frac{b}{a}$ is positive, hence, $\frac{b}{a}$ is negative.

Since we know that a is negative, b must be positive.

Thus, we have $a < 0, b > 0, \& \ c < 0$.

The correct answer is Option D.

194. From the graph given below, when $x = 0$, the value of y i.e. $f(0)$ is negative since the graph intersects the Y-axis below the origin.

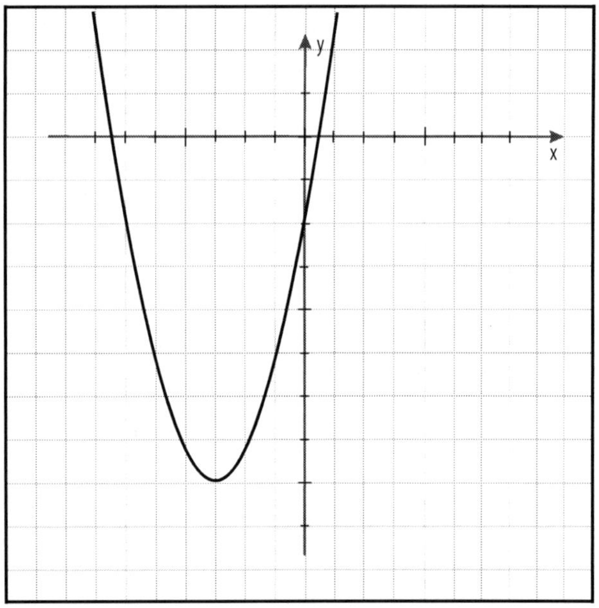

Substituting $x = 0$ in $f(x)$, we get $f(0) = 0 + 0 + c = c$. Thus, the value of c must be negative.

Again, the graph intersects the X-axis at two points (the roots), one on the positive X-axis and the other on the negative X-axis. Hence, product of the roots is negative.

We know that the product of the roots of the quadratic function:

$f(x) = ax^2 + bx + c$ is given by $\frac{c}{a}$.

Thus, we can say that $\frac{c}{a}$ is negative.

Since we know that c is negative, a must be positive.

Now, one root is negative and the other is positive. However, the negative root is bigger in magnitude than the positive root. Thus, the sum of the roots must be negative.

We know that the sum of the roots of the quadratic function: $f(x) = ax^2 + bx + c$ is given by $-\frac{b}{a}$.

Thus, we can say that $-\frac{b}{a}$ is negative, hence, $\frac{b}{a}$ is positive.

Since we know that a is positive, b must be positive as well.

Thus, we have $a > 0, b > 0,\ \&\ c < 0$.

The correct answer is Option C.

195. From the graph given below, the intercept on the y axis is the value of $f(x)$ at $x = 0$ i.e. $f(0)$.

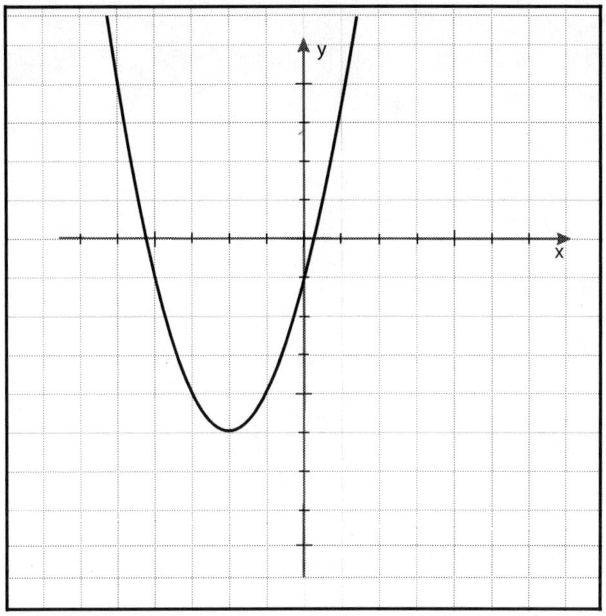

We can see that $f(0) = a(0)^2 + b(0) + c = c$.

Thus, the intercept on the y axis is equal to c (Note: we can see that the graph intersects the y axis below the origin. This implies that the value of c is negative or y-intercept $= -|c|$.

The correct answer is Option D.

196. We have $f(x) = x^2 - 12x + 20$.

Since the coefficient of x^2 is positive, the graph must be open upwards.

Hence, Option B is not possible.

Again, since the value of $f(0) = 20$ (a positive number), it implies that the graph would intersect the Y-axis above the origin.

Hence, Option D is not possible.

In a quadratic $ax^2 + bx + c$, the sum of roots is given by $-\frac{b}{a}$.

In the above quadratic, the sum of roots would be $[-(\frac{-12}{1})] = 12$.

Thus, the sum of roots would be positive.

In Option C, both roots are negative (since the graph intersects the X-axis to the left of the origin) implying that the sum of roots should also be negative.

Hence, Option C is not possible.

Thus, Option A, which satisfies all of the above conditions, must be the answer.

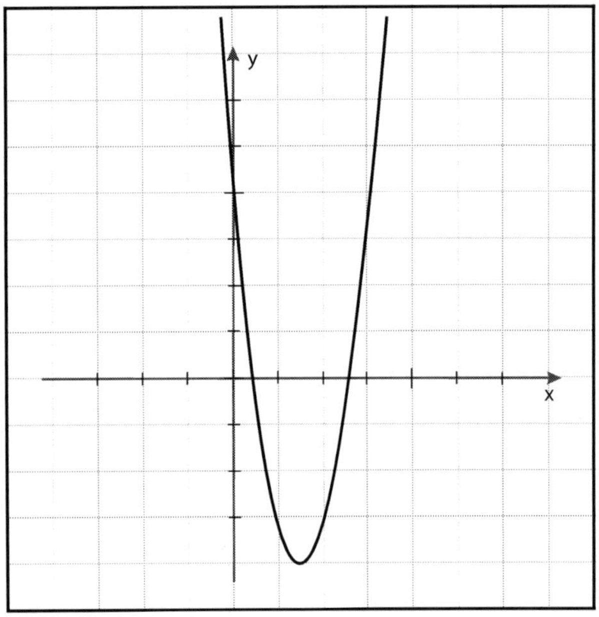

The correct answer is Option A.

Alternate approach:

We have $f(x) = x^2 - 12x + 20$.

Since the value of $f(0) = +20$, Option D is not possible.

At $y = f(x) = 0$, we have $x^2 - 12x + 20 = 0 => x^2 - 2x - 10x + 20 = 0$

$=> x(x-2) - 10(x-2) = 0 => (x-2)(x-10) = 0$

=> $x = +2$ or $+ 10$ (Both roots are positive)

Only the graph in Option A shows that both roots of the quadratic $f(x) = x^2 - 12x + 20$ are positive.

197. We have $f(x) = x^2 + 6x + 9$.

Since the coefficient of x^2 is positive, the graph must be open upwards.

Hence, Option C is not possible.

Again, since the value of $f(0) = 9$, it implies that the graph would intersect the Y-axis above the origin.

Hence, options A and B are not possible.

Thus, Option D, which satisfies all of the above conditions, must be the answer.

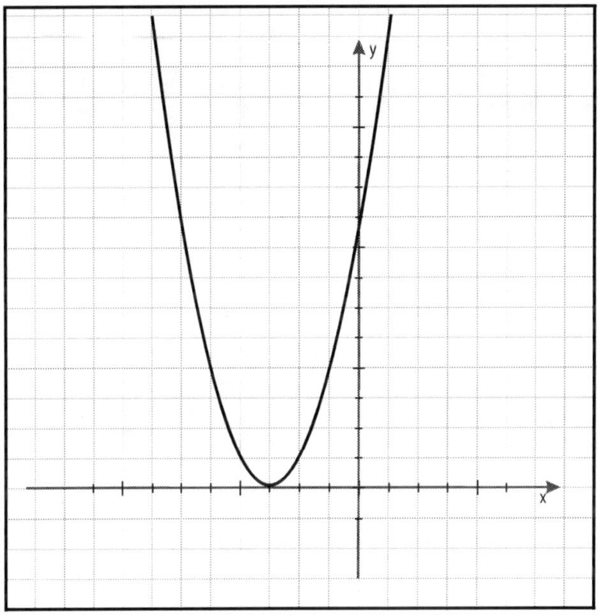

The correct answer is Option D.

198. We see that the graph of $f(x)$ is open upwards.

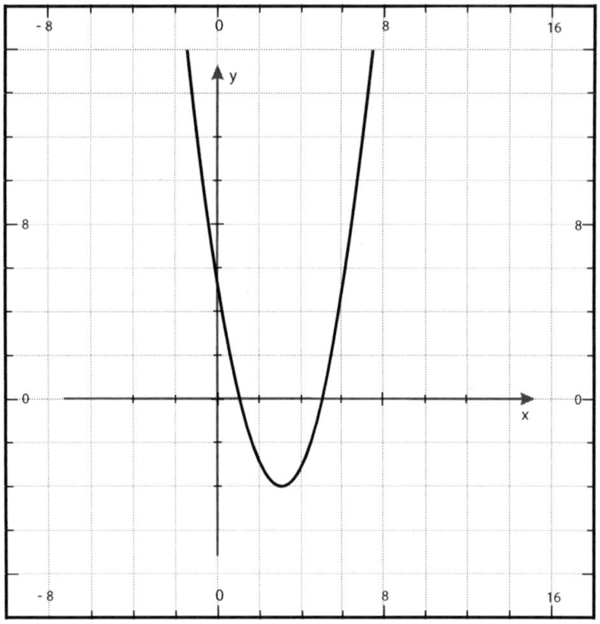

$y = f(x)$

Thus, the coefficient of x^2 must be positive.

Thus, Option D cannot be the solution.

Again, we can observe that the graph of $f(x)$ intersects the Y-axis above the origin at the point between $(0, 4)$ and $(0, 6)$, i.e. the value of $f(0)$ is positive.

In Option A, $f(0) = 8$. Thus, though $f(0)$ is positive, the value of $f(0)$ does not fall in the above range.

Hence, Option A cannot be the solution.

In Option B, $f(0) = 6$. Again, though $f(0)$ is positive, the value of $f(0)$ does not match the above result (i.e. between 4 and 6).

Thus, Option C must be the answer. Let us verify.

$f(0) = (0 - 3)^2 - 4 = 9 - 4 = 5.$

Thus, $f(0)$ is positive and also lies between the values 4 and 6.

The correct answer is Option C.

199. We have $f(x) = x^2 + (a^2 - b)x + b^2 - a^2$.

Since $a > 1 \Rightarrow a^2 > 1$, while $0 < b < 1$, we can say that $a^2 - b > 0$.

Also, since $0 < b < 1 \Rightarrow 0 < b^2 < 1$, we can say that $b^2 - a^2 < 0$.

Thus, we can say that the coefficient of x i.e. $(a^2 - b)$ is positive while the constant term i.e. $(b^2 - a^2)$ is negative.

So, we have $f(0) = b^2 - a^2 < 0$ i.e. the graph of $f(x)$ intersects the Y-axis below the origin.

The product of the roots is given by $\frac{\text{(Constant term)}}{\text{(Coefficient of } x^2)} = \frac{(b^2-a^2)}{1} < 0$. Thus, for $f(x)$, the product of the roots is negative.

Also, the sum of the roots is given by $-\frac{\text{Coefficient of } x}{\text{Coefficient of } x^2} = -(\frac{a^2-b}{1}) = -(a^2 - b) < 0$; [since $(a^2 - b)$ is positive].

Thus, for $f(x)$, the sum of the roots will be negative.

Since the sum of roots and the product of the roots are negative, we must have one positive root and one negative root, with the negative root having a greater magnitude than the positive root.

From the above conditions, we can say that

Option A is not possible since both roots are shown positive, also $f(0)$ is shown positive.

Option B is not possible since the negative root is smaller in magnitude than the positive root implying that the sum of roots will be positive.

Option C is not possible since the value of $f(0)$ is shown as positive.

Thus, Option D must be correct.

In Option D, the value of $f(0)$ is negative. Also, the magnitude of the negative root is greater than that of the positive root implying that the sum of roots will be negative.

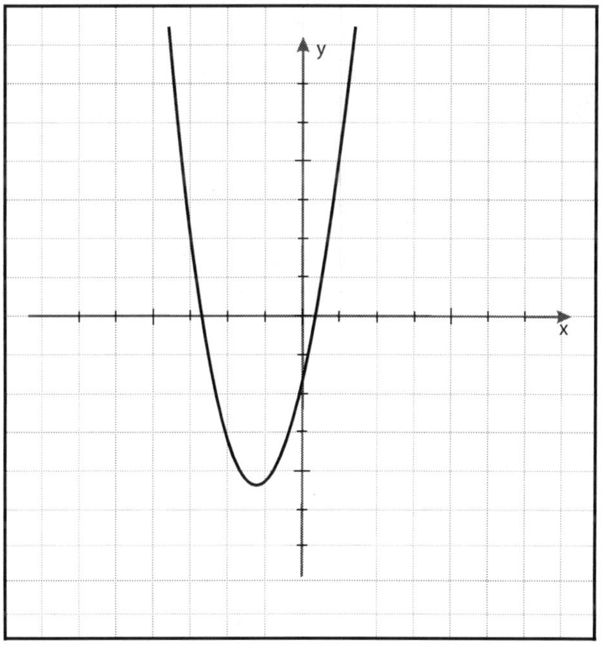

The correct answer is Option D.

200. we have $f(x) = -x^2 + (a - b)x + b^2$

Since the coefficient of x^2 is negative, the graph of $f(x)$ must be open downwards.

Also, since $0 < a < 1$ while $3 < b < 5$, we can say that $a - b < 0$ (in fact, $(a - b)$ lies between $1 - 3 = -2$ and $0 - 5 = -5$. Thus, we have $-5 < a - b < -2$).

Also, b^2 is positive and the value of b^2 must lie between 9 and 25 i.e. $9 < b^2 < 25$ (since $3 < b < 5$).

Thus, we can say that the coefficient of x i.e. $(a - b)$ is negative while the constant term i.e. b^2 is positive.

So, we have $f(0) = b^2 > 0$ i.e. the graph of $f(x)$ intersects the Y-axis above the origin at a value between 9 and 25.

The product of the roots is given by $\frac{\text{Constant term}}{\text{Coefficient of } x^2} = \frac{b^2}{-1} < 0$. Thus, for $f(x)$, the product of the roots is negative.

Also, the sum of the roots is given by $-\frac{\text{coefficient of } x}{\text{coefficient of } x^2} = -\left(\frac{a-b}{-1}\right) = a - b < 0$ [since $(a - b)$ is negative].

Thus, for $f(x)$, the sum of the roots will be negative.

Since the sum of roots and the product of the roots are negative, we must have one positive root and one negative root, with the negative root having a greater magnitude than the positive root.

From the above conditions, we can say that

Option A is not possible since the graph of $f(x)$ is open upwards.

Option B is not possible since the value of $f(0)$ is shown as negative.

Option C is not possible though $f(0)$ is positive since the value of $f(0) = 4$ (we had concluded that $f(0) = b^2$ and $9 < b^2 < 25$).

Thus, Option D must be correct.

In Option D, the value of $f(0)$ is positive and lies between 9 and 25, in fact $f(0) = 18$. Also, the magnitude of the negative root is greater than that of the positive root, implying that the sum of roots will be negative (in fact, the roots are 3 and -6 implying that the sum of the roots is -3; which satisfies the condition that the sum of the roots, i.e. $(a - b)$ must lie between -2 and -5).

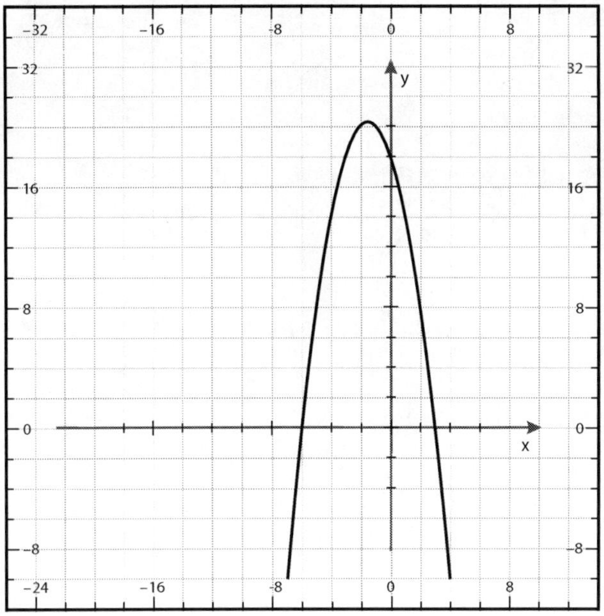

The correct answer is Option D.

201. From the graph, it is quite clear that $f(x) = x^2 + 2x - 8$ refers to the parabola and $g(x) = 2x + 8$ refers to the straight line.

From the above graph, it can be seen that $f(x)$ and $g(x)$ intersect at two points whose coordinates are given by $(-4, 0)$ and $(4, 16)$.

Let us consider the region to the left of $(-4, 0)$ and the region to the right of $(4, 16)$:

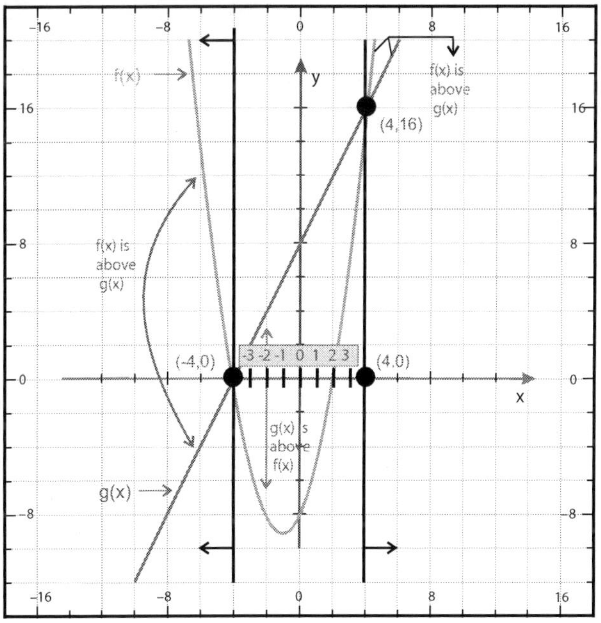

In the above diagram, we see that in both regions, i.e. to the left of $(-4, 0)$ and to the right of $(4, 16)$, the graph of $f(x)$ (i.e. the curve) is placed above the graph of $g(x)$ (i.e. the straight line).

Thus, $f(x) > g(x)$.

However, in the region in between the points of intersection of the graphs, i.e. between $(-4, 0)$ and $(4, 16)$, the graph of $g(x)$ is placed above the graph of $f(x)$.

Thus, $g(x) > f(x)$.

Thus, the integer values of x where $g(x) > f(x)$ is given by: $-3, -2, -1, 0, 1, 2$ and 3. The values of $x = -4$ & $+4$ are excluded as at these values of x, $g(x) = f(x)$ (as shown by the points encircled in black in the above image).

Thus, there are seven integer values of x satisfying $g(x) > f(x)$.

The correct answer is 7.

202. The graph represents $f(x) = x^2 + ax + b$.

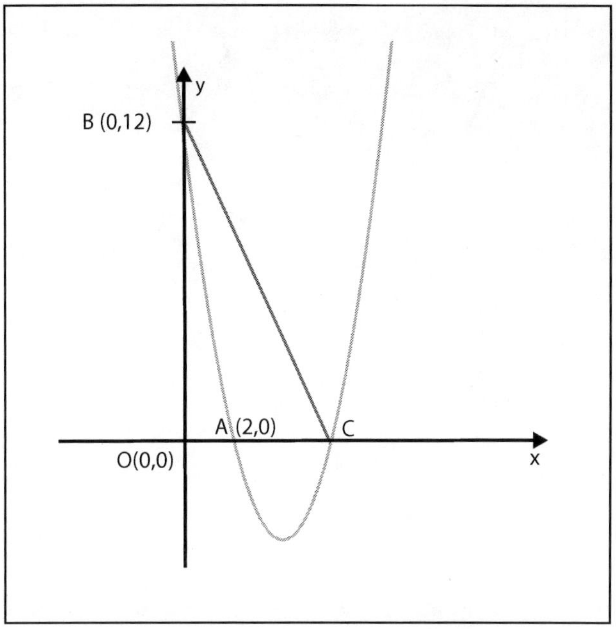

We can see that $f(0)$ is given in the graph as 12 (i.e. the point where $f(x)$ intersects the Y-axis).

Thus, the length of OB is 12 units.

Thus, we have $b = 12$.

In a quadratic, the product of the roots is given as $\frac{\text{Constant term}}{\text{Coefficient of } x^2} = \frac{b}{1} = b = 12$.

We can see that one root is given by the point A i.e. $(2, 0)$, implying that one root of the quadratic is 2.

Since the product of the roots is 12, we can say that the other root must be $\frac{12}{2} = 6$.

Thus, coordinates of point C are $(6, 0)$.

Thus, the length of OC is 6 units.

Hence, area of $\triangle$BOC $= \frac{1}{2} \times \text{base} \times \text{height} = \frac{1}{2} \times (\text{OC}) \times (\text{OB}) = \frac{1}{2} \times 6 \times 12 = 36$ square units.

The correct answer is 36.

5.3.3 Advanced equations

203. Squaring both sides of the given equation, we have

$(\sqrt{2x-8})^2 = (\sqrt{x+5}-3)^2$

$=> 2x - 8 = (\sqrt{x+5})^2 - 2(\sqrt{x+5}) \times 3 + 3^2$

$=> 2x - 8 = x + 5 + 9 - 6\sqrt{x+5}$

$=> 6\sqrt{x+5} = 14 + x - 2x + 8$

$=> 6\sqrt{x+5} = 22 - x$

Squaring both sides of the above equation

$(6\sqrt{x+5})^2 = (22-x)^2$

$=> 36(x+5) = 484 + x^2 - 44x$

$=> x^2 - 44x - 36x + 484 - 180 = 0$

$=> x^2 - 80x + 304 = 0$

$=> x^2 - 4x - 76x + 304 = 0$

$=> x(x-4) - 76(x-4) = 0$

$=> (x-4)(x-76) = 0$

$=> x = 4$ or 76

Now, since squaring an equation can result in extraneous solutions, we need to verify if both the above values of x satisfy the original equation.

For $x = 4$:

LHS $= \sqrt{2x-8} = \sqrt{2 \times 4 - 8} = 0.$

RHS $= \sqrt{x+5} - 3 = \sqrt{4+5} - 3 = 3 - 3 = 0.$

Hence, LHS = RHS.

Thus, $x = 4$ is a correct solution.

For $x = 76$:

LHS $= \sqrt{2x-8} = \sqrt{2 \times 76 - 8} = \sqrt{144} = 12.$

RHS $= \sqrt{x+5} - 3 = \sqrt{76+5} - 3 = 9 - 3 = 6.$

Hence, LHS > RHS.

Thus, $x = 76$ is not a correct solution.

Hence, we have $x = 4$.

The correct answer is Option A.

The above is a lengthy and time-consuming process; a better approach would be to plug-in option values in the given equation. We can see that if $x = 4$, the equation is satisfied and hence, it is the solution.

204. In order to solve this equation, we need to equate the exponents on either side after having reduced either base to the same value.

Thus, $(2^{x+2})^{x+6} = 4^6$

$=> 2^{(x+2)(x+6)} = (2^2)^6$

$=> 2^{(x+2)(x+6)} = 2^{12}$

Equating the exponents on either side:

$(x + 2)(x + 6) = 12$

$=> x^2 + 8x + 12 = 12$

$=> x^2 + 8x = 0$

$=> x(x + 8) = 0$

$=> x = 0$ or -8.

Since x is non-zero, we have $x = -8$.

The correct answer is option A.

The above is a lengthy and time-consuming process; a better approach would be to plug-in option values in the given equation. We can see that if $x = -8$, the equation is satisfied and hence, is the solution.

205. In order to solve this equation, we need to equate the exponents on either side after having reduced either base to the same value.

Thus, $3^{2x-1} = 3^5$

Equating the exponents on either side:

$2x - 1 = 5 => 2x = 6 => x = 3$.

The correct answer is 3.

206. Keeping the radical term on the left and transferring the other two terms to the right, we have

$\sqrt{3x - 8} = x - 2$

Squaring both sides of the above equation

$(\sqrt{3x - 8})^2 = (x - 2)^2$

$=> 3x - 8 = x^2 - 4x + 4$

$=> x^2 - 4x - 3x + 4 + 8 = 0$

$=> x^2 - 4x - 3x + 12 = 0$

$=> x(x - 4) - 3(x - 4) = 0$

$=> (x - 4)(x - 3) = 0$

$=> x = 4$ or 3.

Now, since squaring an equation can result in extraneous solutions, we need to verify if both the above values of x satisfy the original equation.

At $x = 4$:

LHS $= \sqrt{3x - 8} - x + 2 = \sqrt{3 \times 4 - 8} - 4 + 2 = 2 - 4 + 2 = 0 =$ RHS.

Hence, LHS = RHS.

Thus, $x = 4$ is a correct solution.

At $x = 3$:

LHS $= \sqrt{3x - 8} - x + 2 = \sqrt{3 \times 3 - 8} - 3 + 2 = 1 - 3 + 2 = 0 =$ RHS.

Hence, LHS = RHS.

Thus, $x = 3$ is a correct solution.

Hence, we have both $x = 4$ as well as $x = 3$ as possible solutions.

Thus, absolute difference between the values of x is given by: (the larger value) − (the smaller value)

$= 4 - 3 = 1$.

The correct answer is Option B.

Note: You cannot solve this question by plugging-in values from the options as the question does not ask for the roots of the equation.

207. In order to solve this equation, take the square root on both sides.

Thus, $\sqrt{(x^2 - 17)^2} = \pm\sqrt{64}$

$=> x^2 - 17 = \pm 8$

$=> x^2 - 17 = 8$ or $x^2 - 17 = -8$

From the first equation: $x^2 - 17 = 8 => x^2 = 17 + 8 => x^2 = 25 => x = \pm 5$.

From the second equation: $x^2 - 17 = -8 => x^2 = 17 - 8 => x^2 = 9 => x = \pm 3$.

Thus, the values of x are $5, -5, 3 \& -3$.

Thus, there are four values of x.

The correct answer is Option D.

Note: You cannot solve this question by plugging-in values from the options as the question does not ask for the roots of the equation.

208. Squaring both sides of the given equation, we have

$(\sqrt{x-8})^2 = (\sqrt{x+8} - 2)^2$

$=> x - 8 = (x+8) - 2 \times (\sqrt{x+8}) \times 2 + 4$

$=> x - 8 = x + 8 - 4\sqrt{x+8} + 4$

Keeping the radical term on the left and transferring the other two terms to the right, we have

$4\sqrt{x+8} = x + 8 + 4 - x + 8$

$=> 4\sqrt{x+8} = 20$

$=> \sqrt{x+8} = 5$

Squaring both sides of the above equation, we have

$x + 8 = 25$

$=> x = 17$

The correct answer is 17.

209. Squaring both sides of the given equation:

$\left\{\sqrt{(-2)^{x-5}}\right\}^2 = \left\{(2)^{(x-6)}\right\}^2$

$=> (-2)^{x-5} = (2)^{2(x-6)}$

$=> (-2)^{x-5} = (2)^{2x-12}$

$=> \{(-1) \times (2)\}^{x-5} = 2^{2x-12}$

$=> (-1)^{x-5} \times 2^{x-5} = 2^{2x-12}$

$=> (-1)^{x-5} = \frac{2^{2x-12}}{2^{x-5}}$

$=> (-1)^{x-5} = 2^{(2x-12)-(x-5)}$

$=> (-1)^{x-5} = 2^{x-7} \ldots$(i)

In the above equation, the LHS will take the value of $+1$ or -1 depending on whether the exponent $(x - 5)$ is even or odd respectively. However, the RHS is 2 raised to an exponent, making it positive. Hence, the value of the RHS must be $+1$; thereby the value of $(x - 5)$ must be even.

Thus, since $(x - 5)$ is even, we have from (i):
$1 = 2^{x-7} => 2^0 = 2^{x-7} => x - 7 = 0 => x = 7$.

In such a case, the value of $x - 5 = 7 - 5 = 2$, which is even, as was required.

The correct answer is Option B.

The above is a lengthy and time-consuming process; a better approach would be to plug-in option values in the given equation.

So, if we plug-in option values in the given equation using $x = 4,\ 7,\ 8,\ \ldots$, we find that at $x = 7$, the expression holds true.

At $x = 7$,

LHS = $\sqrt{(-2)^{x-5}} = \sqrt{(-2)^{7-5}} = \sqrt{(-2)^2} = \sqrt{4} = 2$ (since "$\sqrt{}$" returns only the positive square root).

RHS = $2^{(x-6)} = 2^{7-6} = 2$.

Thus, the equation is satisfied since LHS = RHS.

5.3.4 Application of functions

210. Let us analyze the function $h(t)$.

$h(t) = -t^2 + 12t + 30 = -(t^2 - 12t) + 30 = -\{(t - 6)^2 - 36\} + 30 = -(t - 6)^2 + 66 \ldots$ (i)

Since $-(t - 6)^2$ is the negative of a perfect square, it is always non-positive.

Hence, for $h(t)$ to have the maximum value, the value of $(t - 6)^2$ should be minimum. Thus, it should be zero.

Thus, $(t - 6)^2 = 0 => t = 6$.

Thus, the bullet would reach the maximum height above the ground 6 seconds after the pistol was fired.

The maximum height attained can be obtained by substituting $t = 6$ in (i) giving the maximum height as 66 ft.

The correct answer is Option D.

211. Since the height above the ground is 57 feet, we have

Given that $h(t) = -t^2 + 12t + 30 = 57$

$=> -t^2 + 12t - 27 = 0$

$=> t^2 - 12t + 27 = 0 => t^2 - 9t - 3t + 27 = 0$

$=> t(t - 9) - 3(t - 9) = 0$

$=> (t - 3)(t - 9) = 0 => t = 3$ or 9.

Since we want the earliest time, we have $t = 3$ i.e. after three seconds.

The correct answer is Option B.

Alternatively, you can also get the answer by plugging in the option values in the function $h(t)$ and get the answer.

212. The height of the bullet above the ground when it was fired is given by substituting $t = 0$ in $h(t)$.

Thus, $h(0) = -0 + 0 + 30 = 30$.

We need to find the time when the height of the bullet above the ground will again be 30 feet.

Thus, $h(t) = 30$

$=> -t^2 + 12t + 30 = 30$

$=> -t^2 + 12t = 0$

$=> -t(t - 12) = 0 => t = 0$ or 12.

Here, $t = 0$ refers to the time when the bullet was fired.
Thus, we need the other value of t.

Hence, $t = 12$ i.e. the bullet will again reach a height of 30 feet after 12 seconds.

The correct answer is Option C.

213. Given that the total cost of manufacturing x vases = $\$(x^2 - 17x + 80)$.

Total sales proceeds from the vases = $\$25x$.

Hence, profit = total sales − total cost

$= \$(25x - (x^2 - 17x + 80)) = \$(-x^2 + 42x - 80)$

Since there is neither profit nor loss, we have $-x^2 + 42x - 80 = 0$

$=> x^2 - 42x + 80 = 0$

$=> x^2 - 40x - 2x + 80 = 0$

$=> x(x - 40) - 2(x - 40) = 0$

$=> (x - 40)(x - 2) = 0$

$=> x = 40$ or 2.

Hence, for neither profit nor loss, the company needs to manufacture either 2 or 40 vases.

The correct answer is Option B.

214. From the last question, we know that the profit made by selling x vases is given by $\$(-x^2 + 42x - 80)$.

We need to find for what value of x the above profit will become the maximum.

Thus, we have

$-x^2 + 42x - 80 = -(x^2 - 42x) - 80 = -\{(x^2 - 2 \times x \times 21 + 21^2) - 21^2\} - 80$

$= -(x - 21)^2 + 21^2 - 80$

$= -(x - 21)^2 + 361$

Thus, the profit will be maximum when the term $\{-(x - 21)^2\}$, which is the negative of a perfect square, becomes zero.

This happens at $x = 21$ and the maximum value of the profit becomes \$361.

Thus, the company needs to manufacture 21 vases in order to maximize the profit.

The correct answer is Option D.

215. The number of bacteria at the start of the experiment = N.

Since the number of bacteria trebles every hour, the number of bacteria at the end of every hour is shown in the table below:

At the end of	Number of bacteria
1 hour	$3 \times N$
2 hours	$3 \times 3N = 3^2 \times N$
3 hours	$3 \times 3^2 \times N = 3^3 \times N$
4 hours	$3 \times 3^3 \times N = 3^4 \times N$
...	...
t hours	$3^t \times N$

Thus, we see that at the end of t hours, the number of bacteria = $3^t \times N$.

The correct answer is Option D.

216. Since the bacteria count trebles every hour, an hour before, the count would have been $1/3^{rd}$ of what is now i.e. 72900.

Since 72900 is the count after 6 hours, the count would have been $\frac{1}{3}^{rd}$ after $(6 - 1) = 5$ hours.

Thus, the number of bacteria was one-third of the final number five hours after the start of the experiment.

The correct answer is Option D.

217. We know that the number of bacteria at the end of t hours = $3^t \times N$.

We know that at the end of six hours, the number of bacteria present = 72,900.

Thus, we have $3^6 \times N = 72,900 => N = \frac{72,900}{3^6} = \frac{72,900}{729} = 1,000$.

The correct answer is Option C.

218. We have $v(t) = -t^2 + 20t + 100 = -(t^2 - 20t) + 100$

$= -\{(t^2 - 2 \times t \times 10 + 10^2) - 10^2\} + 100 = -(t - 10)^2 + 200.$

The number of vehicles on the street is the maximum if the value of the term $\{-(t - 10)^2\}$ is zero (since the term is the negative of a perfect square, it tends to reduce the value of the function).

This is possible at $t = 10$ i.e. 10 hours past midnight (the maximum no. of vehicles is 200).

The correct answer is 10.

219. We have $v(t) = -t^2 + 20t + 100$.

The number of cars at midnight is given by $v(0) = -0 + 0 + 100 = 100$.

Thus, we need to find at what other time the number of cars would also be 100.

Thus, we have $-t^2 + 20t + 100 = 100$

$=> -t^2 + 20t = 0 => -t(t - 20) = 0$

$=> t = 0$ or 20.

Here, $t = 0$ refers to midnight.

Hence, the other time is at $t = 20$ i.e. 20 hours past midnight.

The correct answer is 20.

5.4 Geometry & Trigonometry

5.4.1 Geometry

220. We have ∠BOE = ∠AOD (vertically opposite angles).

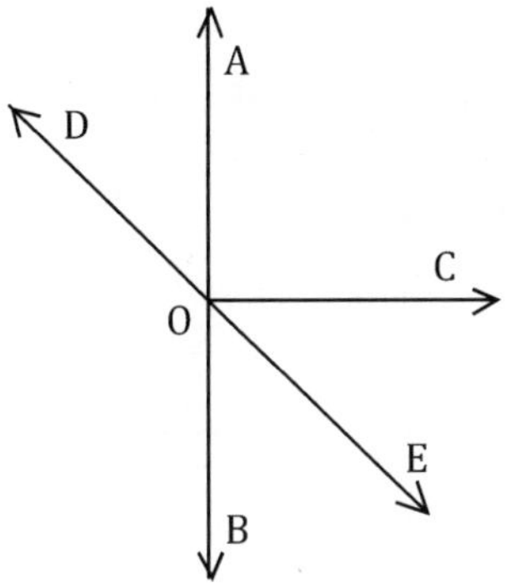

Since ∠AOD = 2 × ∠COE => ∠BOE = 2 × ∠COE.

Let ∠COE = $x°$.

Thus, ∠BOE = $2x°$.

Since OC is perpendicular to AB, we have ∠COB = ∠COA = 90°.

Thus, ∠COB = ∠COE + ∠BOE = 90° => $x + 2x = 90°$ => $x = 30°$.

Thus, ∠COE = 30°.

Hence, ∠EOA = ∠COE + ∠COA = 30° + 90° = 120°.

The correct answer is Option C.

221. ∠MPD = ∠BPO and ∠ABC = ∠OBP (vertically opposite angles).

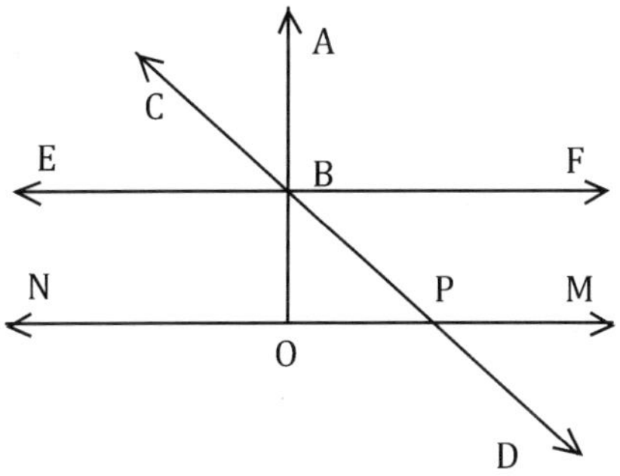

Since ∠MPD = ∠ABC + 20° => ∠BPO = ∠OBP + 20°.

Let ∠OBP = $x°$ => ∠BPO = $x° + 20°$.

We know that in a triangle, the sum of angles is 180°.

In triangle BOP, ∠BOP = 90° (since BO is perpendicular to OP).

Thus, $\angle BPO + \angle OBP = 180^\circ - 90^\circ = 90^\circ \Rightarrow x + 20^\circ + x = 90^\circ \Rightarrow x = 35^\circ$.

Thus, $\angle BPO = 35^\circ + 20^\circ = 55^\circ$.

$\angle CBE = \angle FBP$ (vertically opposite angles).

Also, $\angle FBP = \angle BPO$ (alternate angles).

Thus, $\angle CBE = \angle BPO = 55^\circ$.

The correct answer is Option C.

222. Area of the square ABCD = $(\text{side})^2 = AB^2 = 2^2 = 4$ square units.

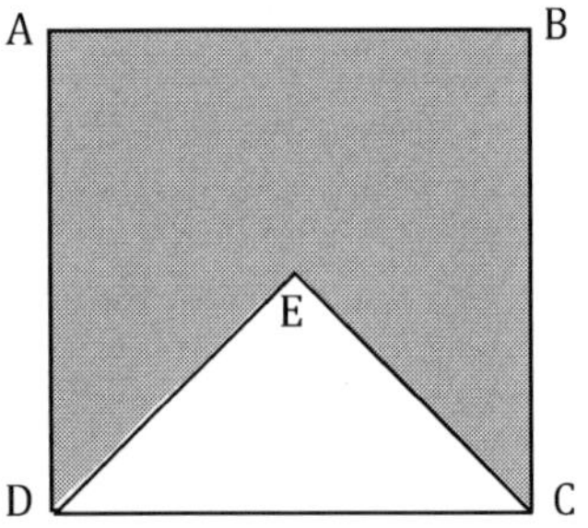

Area of the triangle DEC

$= \frac{1}{2} \times \text{base} \times \text{height} = \frac{1}{2} \times DC \times \text{height} = \frac{1}{2} \times 2 \times 1 = 1$ square units.

Hence, the area of the shaded region

= area of square ABCD – area of triangle DEC = $4 - 1 = 3$ square units.

The correct answer is 3.

223. In a parallelogram, the opposite sides are equal.

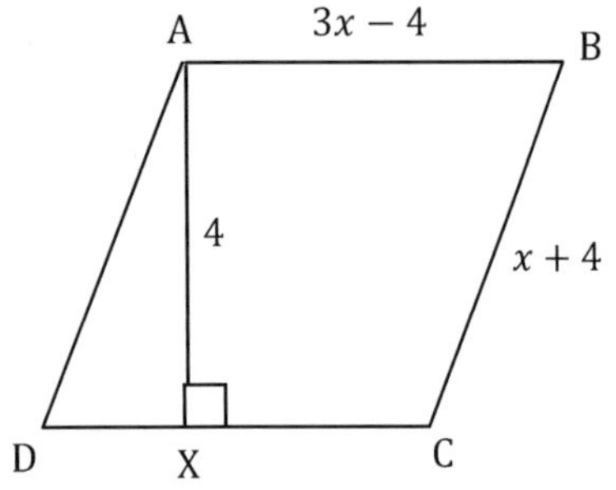

Thus, AB = CD and BC = AD.

Thus, perimeter of ABCD = AB + BC + CD + DA = 2 × (AB + BC)
$= 2((3x - 4) + (x + 4)) = 8x$.

Thus, we have $8x = 24 \Rightarrow x = 3$ units.

Hence AB = CD = $3x - 4 = 3 \times 3 - 4 = 5$ units.

Thus, the area of the parallelogram ABCD = base × height = CD × AX = 5 × 4 = 20 square unit.

The correct answer is 20.

224. When the rectangle is folded such that AD and BC overlap, we form a cylinder whose base perimeter (perimeter of base circle) is AB and height is BC.

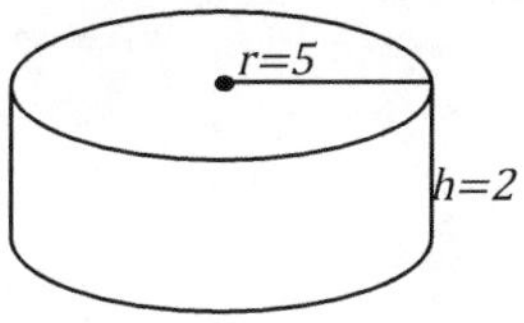

Let the radius of the cylinder be r.

Thus, we have $2\pi r = 10\pi \Rightarrow r = 5$ units.

Also, height of the cylinder = h = 2 units.

Hence, volume of the cylinder = $\pi r^2 h = \pi \times 5^2 \times 2 = 50\pi$ cubic units.

The correct answer is option C.

225. The approach to solve this question is by equating the volume of both the solids i.e. 'Volume of the prism = Volume of all cubes'

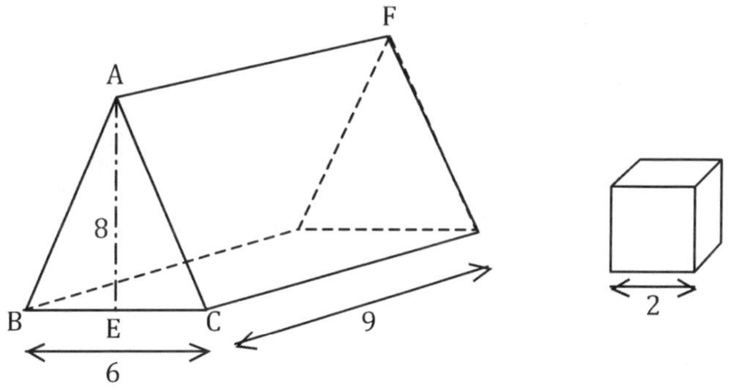

The volume of the prism = area of the base × length = (area of △ABC) × AF

$= \left(\frac{1}{2} \times BC \times AE\right) \times AF = \frac{1}{2} \times 6 \times 8 \times 9 = 216$ cubic units.

Volume of one cube = $(\text{side})^3 = 2^3 = 8$ cubic units.

Hence, number of cubes possible = $\frac{\text{Volume of the prism}}{\text{Volume of one cube}} = \frac{216}{8} = 27$.

The correct answer is Option B.

226. The area of the shaded region = (area of the square) − (area of the circle)

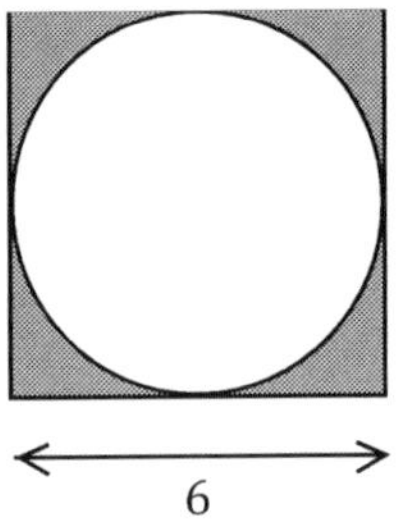

Area of the square = $(\text{side})^2 = 6^2 = 36$ square units.

The diameter of the circle is the same as one side of the square.

Hence, diameter of the circle = 6 units.

Hence, radius of the circle = 3 units.

Hence, area of the circle = $\pi \times (\text{radius})^2 = 9\pi$ square units.

Hence, the required area = $(36 - 9\pi)$ square units = $9(4 - \pi)$ square units.

The correct answer is Option D.

227.

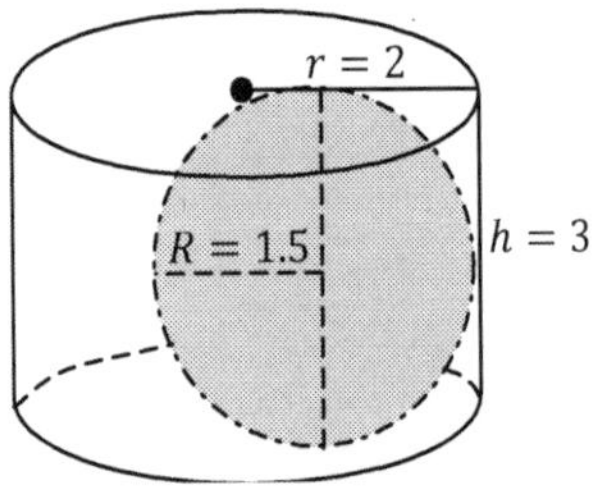

If we choose a sphere of radius equal to that of the cylinder i.e. 2 units, then the diameter of the sphere would be $2 \times 2 = 4$ units.

However, since the height of the cylinder is only 3 units, a part of the sphere would protrude outside the cylinder and hence, would not completely fit inside the cylinder.

Thus, in order that the sphere completely fits inside the cylinder, we need to choose a sphere of diameter 3 units, i.e. radius 1.5 units (the sphere placed inside the cylinder is shown in the diagram above).

Hence, the volume of the sphere having radius 1.5 units = $\frac{4}{3} \times \pi \times (\text{radius})^3 = \frac{4}{3} \times \pi \times (1.5)^3$

$= \frac{4}{3} \times \pi \times \left(\frac{3}{2}\right)^3 = \frac{4}{3} \times \pi \times \frac{3}{2} \times \frac{3}{2} \times \frac{3}{2} = \frac{9\pi}{2}$ cubic units.

The correct answer is Option B.

228. The total surface area of a cube is equivalent to area of 6 identical squares.

Hence, the total surface area = $6 \times (\text{side})^2 = 96k$ => $(\text{side})^2 = \frac{96k}{6} = 16k$

=> side $= \sqrt{16k} = 4\sqrt{k}$

Thus, volume of the cube = $(\text{side})^3 = \left(4\sqrt{k}\right)^3 = 4^3 \times \left(k^{\left(\frac{1}{2}\right)}\right)^3 = 64k^{\left(\frac{3}{2}\right)} = 64k\sqrt{k}$.

The correct answer is Option B.

229. Let the radius of the sphere be r.

Thus, radius of the cylinder = $3r$.

Let the height of the cylinder be h.

Volume of the sphere = $\frac{4}{3} \times \pi \times r^3$.

Volume of the cylinder = $\pi \times (3r)^2 \times h = 9\pi r^2 h$.

Since the volume of the sphere is double that of the cylinder, we have

$\frac{4}{3}\pi r^3 = 2 \times 9\pi r^2 h$ => $\frac{r}{h} = 18 \times \frac{3}{4} = \frac{27}{2}$ => $\frac{h}{r} = \frac{2}{27}$.

However, we need to find the ratio of the height of the cylinder to the radius of the cylinder, i.e. $\frac{h}{3r}$.

Thus, we have $\frac{h}{3r} = \frac{1}{3} \times \frac{h}{r} = \frac{1}{3} \times \frac{2}{27} = \frac{2}{81}$.

Hence, the required ratio is 2 : 81.

The correct answer is Option A.

230.

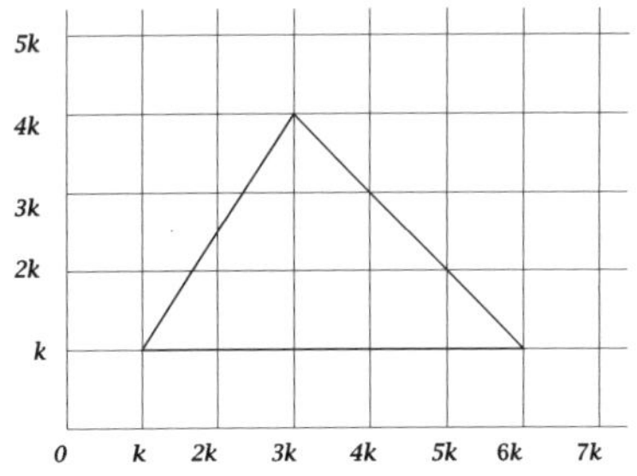

The area, A, of any triangle, is given by the formula

$A = \frac{1}{2}bh$; where h is the length of an altitude corresponding to a side with the length b.

Since $A = 35$, $bh = 70$.

The figure supporting the question shows that $b = 5k$, $h = 3k$, so $bh = 15k^2$.

Substituting $bh = 70$, we have that $15k^2 = 70$.

=> $k^2 = \frac{70}{15} = \frac{14}{3}$.

Thus, $k = \sqrt{\frac{14}{3}} = \sqrt{\frac{14 \times 3}{3 \times 3}} = \frac{\sqrt{42}}{3}$.

The correct answer is Option D.

231. Let us understand what the question asks. It states, *The distance between the line joining the centers of the circles and the line drawn through point P parallel to the line connecting the centers* means that we have to find out the value of PQ; see the diagrams below.

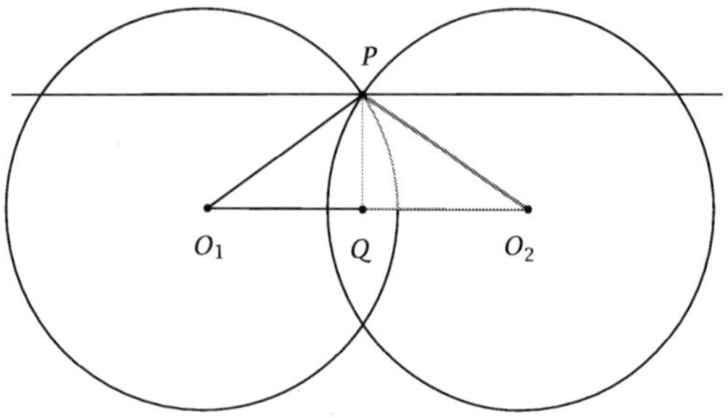

The answer comes from the application of the Pythagorean theorem and right triangles.

First, notice that $\triangle PO_1O_2$ is isosceles, and PQ is the altitude of $\triangle PO_1O_2$ by construction. Therefore, $\triangle PQO_2$ is a right triangle. In an isosceles triangle, an altitude dropped from the vertex divides the base in two equal halves: O_1Q and QO_2. Each half is of length 8 since the halves make up $O_1O_2 = 16$. Now, we can get the length of PQ with the help of the Pythagorean theorem:

$$PQ = \sqrt{(PO_2)^2 - (QO_2)^2}$$
$$= \sqrt{10^2 - 8^2} = 6.$$

The correct answer is Option C.

232. The key is to use the fact that the sum of the angles in a triangle is 180°. Each of the three triangles shown in the diagram is a right triangle. It follows that $x = 180 - 45 - 90 = 45°$.

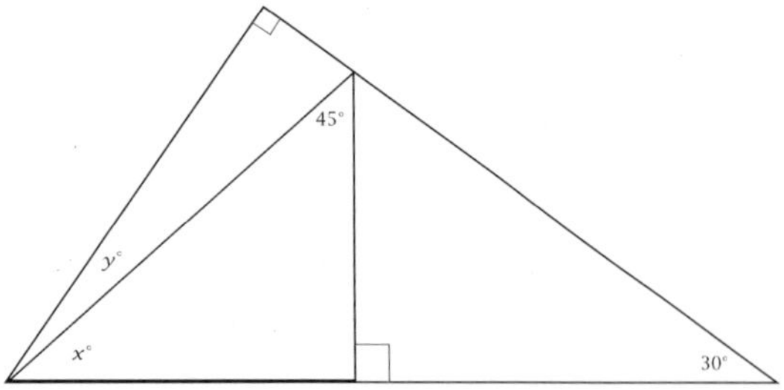

Note: Figure not drawn to scale

For the scalene outermost triangle, we know that $180 = 90+30+(x+y)$, thus $x+y = 60°$. So, $y = 15°$, and $2x - y = 90° - 15° = 75°$.

The correct answer is Option D.

233. The key to this problem is to remember that the sides of a $30° - 60° - 90°$ triangle are related as

$1 : \sqrt{3} : 2$, and the sides of a $45° - 45° - 90°$ triangle are related as $1 : 1 : \sqrt{2}$.

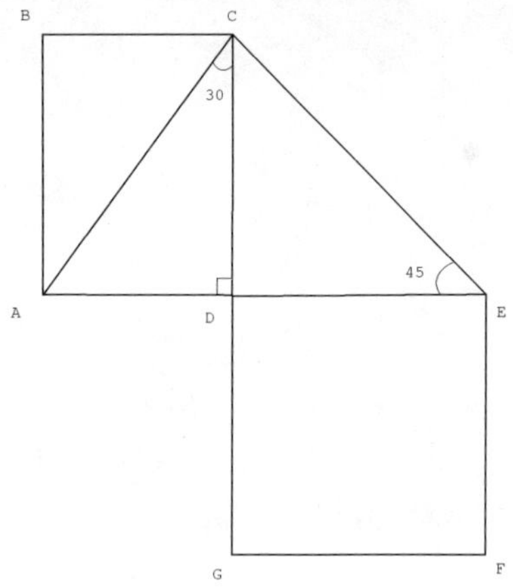

We know that area of $\triangle ACD$ is half of the area of rectangle $ABCD$ thus, area of $\triangle ACD$ = $2\sqrt{3}$.

Again, $\triangle ACD$ is a $30° - 60° - 90°$ triangle, so its sides are related as

$AD : CD : AC :: 1 : \sqrt{3} : 2$.

Hence,

Area of $\triangle ACD = 2\sqrt{3} = \frac{1}{2}AD \times CD = \frac{1}{2}AD \times (\sqrt{3}AD) = \frac{\sqrt{3}}{2}AD^2$

$=> 2\sqrt{3} = \frac{\sqrt{3}}{2}AD^2$

$=> \sqrt{3}AD^2 = 4\sqrt{3} => AD = 2$.

So, $CD = \sqrt{3} \times AD = 2\sqrt{3}$.

From triangle CDE, which is a $45° - 45° - 90°$ triangle, we have $CD = DE$

So, the area of the square $DEFG = DE^2 = CD^2 = (2\sqrt{3})^2 = 12$.

The correct answer is Option D.

234. Let the distance between A and B be a and the distance between B and C be b. Then, $a + b = 150$, and $a = 5b$. To find out b, substitute the latter into the former to obtain that $6b = 150$. So, $b = 25 = BC$.

The correct answer is Option D.

235.

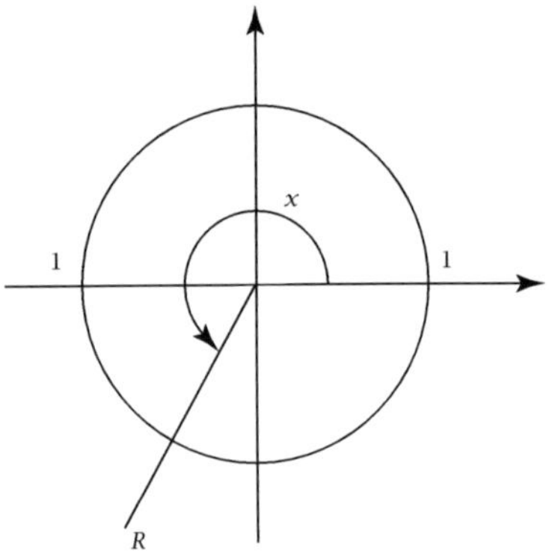

The key is to remember that ray R has completed one revolution around the origin. One revolution with a radius of 1 results in a circle with the circumference of 2π. Then, it is best to proceed by elimination. I is not correct because it states that $D < \frac{3\pi}{2}$, but we know that $D > 2\pi$. Like I, II is not correct because it states that $D < \pi$, and we know that $D > 2\pi$. Now, it is clear that III is the only answer since it is clearly more that 2π.

The correct answer is Option C.

Alternatively, from the diagram, we know that the ray completed one full revolution and after the completion of an additional half revolution but before the three quarter revolution, it stops; thus we can conclude that

$[(2\pi + \frac{1}{2} \times 2\pi) < D < (2\pi + \frac{3}{4} \times 2\pi)]$ OR $[3\pi < D < \frac{7\pi}{2}]$.

236.

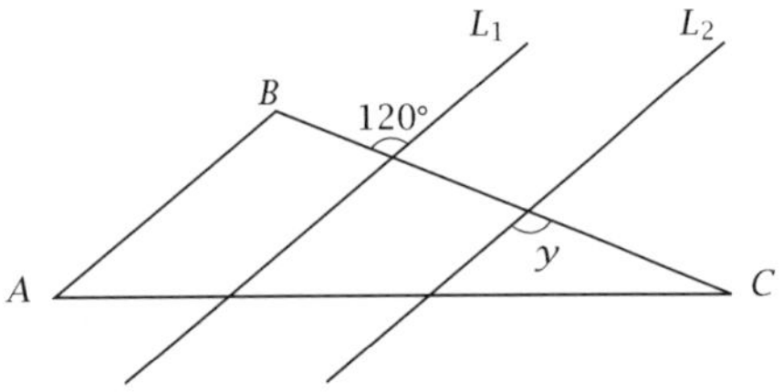

The key is to use the properties of two parallel lines intersected by a transversal. To find the measure of y, we need to relate AB and L_2. Since $AB \| L_1$ & $L_1 \| L_2$, $AB \| L_1 \| L_2$. Thus, the transversal BC generates corresponding angles equal to y; $y = \angle 120°$ since they are vertically opposite angles.

The correct answer is Option D.

237.

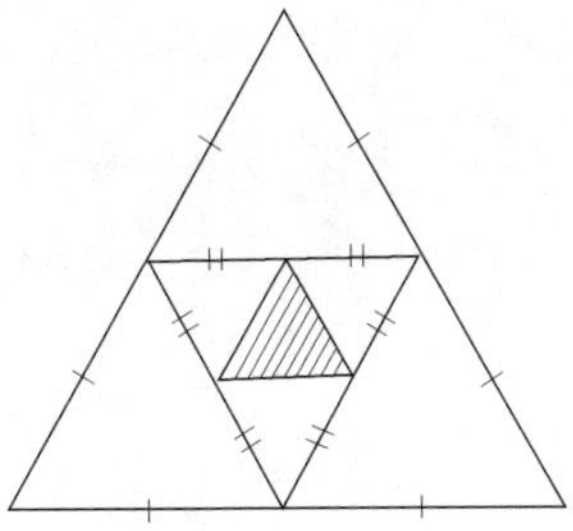

Let the side of the smallest equilateral triangle be s. We know that the area of an equilateral triangle is $\frac{\sqrt{3}}{4}(\text{side})^2$.

Then, its area $= \frac{\sqrt{3}}{4}s^2$.

Since all the triangles are equilateral, the side of the largest equilateral triangle is $4s$.

Its area $= \frac{\sqrt{3}}{4}(4s)^2 = 4\sqrt{3}s^2$.

Dividing the area of the smallest triangle by the largest, we get $\frac{1}{16}$.

For a simple counting argument, see the figure below.

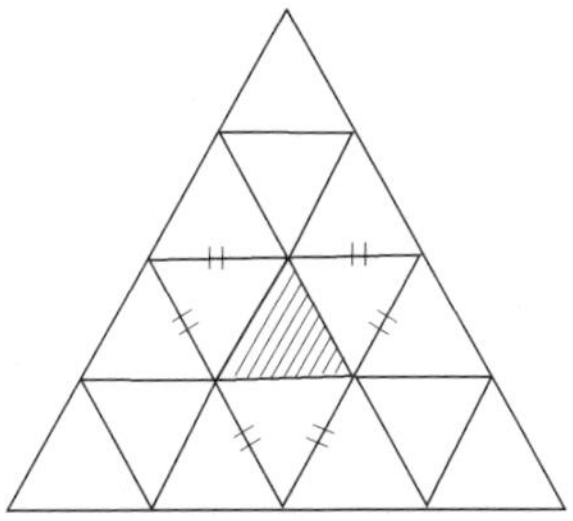

Sixteen equilateral triangles are obtained by connecting the midpoints of the sides of the outer triangles and, then, repeating the same procedures once again for each of the four thus constructed triangles. It follows from this procedure that all 16 triangles are equal by equality of their sides.

The correct answer is Option A.

238. See the figure below.

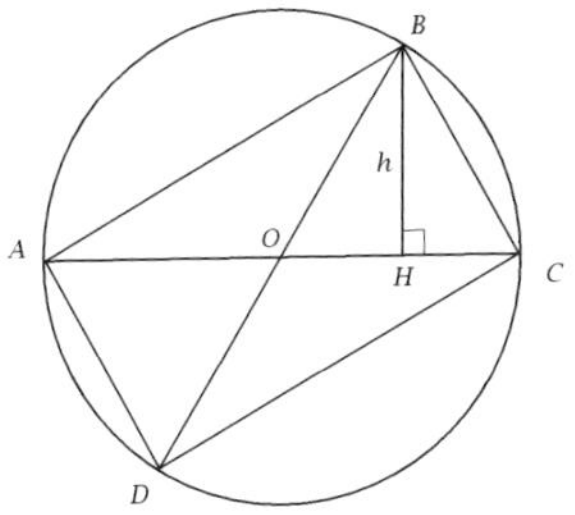

First, note that $\triangle ABC = \triangle CDA$. This follows from the fact that points B, O, and D lie on the same line, which allows to split the above-mentioned triangles into pairs of triangles

$\triangle ABO$ & $\triangle BOC$ and $\triangle AOD$ & $\triangle DOC$ and see that the first triangle in each pair is equal to the second triangle in the other pair, and vice versa. So, it is sufficient to focus on $\triangle ABC$.

The length of the altitude h of $\triangle ABC$ is obtained from the expression $A = \frac{1}{2}bh$ for the area A of $\triangle ABC$. We know that $b = 12$, since b being the base of the triangle and is equal to the diameter of the circle $= 6 \times 2 = 12$.

Thus $A = 18\sqrt{3} = \frac{1}{2} \times 6h \Rightarrow h = (2 \times 18\sqrt{3})/12 = 3\sqrt{3}$.

Then, the area of $\triangle BOC$ is $\frac{1}{2} \times OC \times h = \frac{1}{2} \times 6 \times h = 3 \times 3\sqrt{3} = 9\sqrt{3}$, and

$$\begin{aligned} \boldsymbol{A}(\triangle ODC) &= \boldsymbol{A}(\triangle OBA); \text{where } \boldsymbol{A} \text{ denotes area} \\ &= \boldsymbol{A}(\triangle ABC) - \boldsymbol{A}(\triangle OBC) \\ &= 18\sqrt{3} - 9\sqrt{3} \\ &= 9\sqrt{3}. \end{aligned}$$

The correct answer is Option B.

Alternate approach:

Area of triangle $ABC = \frac{1}{2} \times AC \times H$

Area of triangle $OBC = \frac{1}{2} \times \frac{AC}{2} \times H$

So, area of $OBC = \frac{1}{2} \times$ area of ABC

But area of OBC = area of ODC

$\Rightarrow$ Area of triangle $ODC = \frac{1}{2} \times 18\sqrt{3} = 9\sqrt{3}$

239.

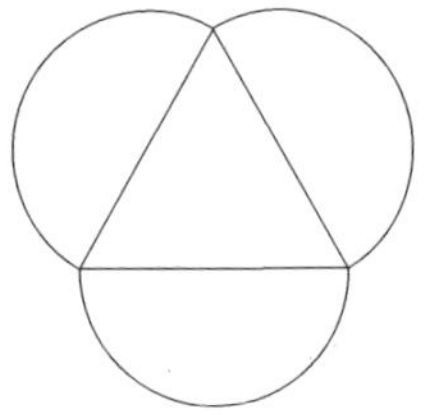

The area of the figure is the area of three semi-circles of radius $40/2 = 20$, plus the area of the equilateral triangle of side 40.

$\Rightarrow$ Area $= [3 \times (\frac{1}{2} \times \pi \times (\frac{40}{2})^2) + \frac{\sqrt{3}}{4} \times 40^2]$

So, the volume of the figure = Height $\times$ Area of the base

Volume $= 5 \times (\frac{3}{2} \times \pi \times 20^2 + \frac{\sqrt{3}}{4} \times 40^2)$

After doing some algebra, the expression above simplifies to $3,000\pi + 2,000\sqrt{3}$.

The correct answer is Option B.

240. Assume that the length and the width of the rectangle are x & y, respectively. See the figure below for notation.

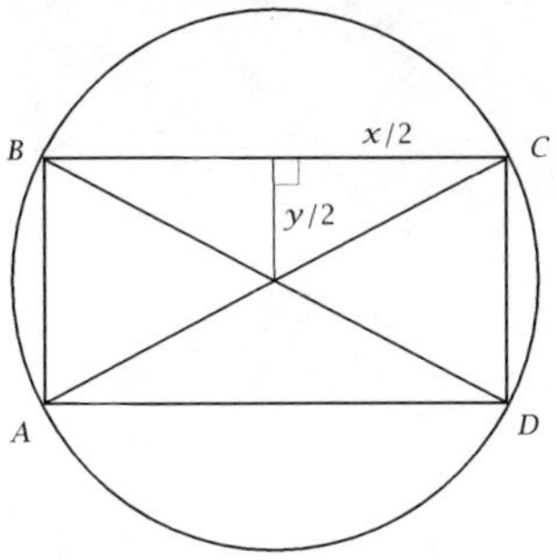

From the figure above, the Pythagorean theorem and the perimeter of $ABCD$ provide us with a system of two equations

$$\left(\frac{x}{2}\right)^2 + \left(\frac{y}{2}\right)^2 = \left(\frac{15}{\sqrt{2}}\right)^2,$$

$$\frac{x^2}{4} + \frac{y^2}{4} = \frac{225}{2},$$

$x^2 + y^2 = 450$ (after simplification).

From the perimeter information, we have that

$2x + 2y = 60,$

$x + y = 30$

$y = 30 - x$

After substituting the value of $y = 30 - x$ into the $x^2 + y^2 = 450$ equation, we have

$x^2 + (30 - x)^2 = 450$, and after simplifying

$x^2 + 30^2 - 2 \times 30 \times x + x^2 = 450$

$=> 2x^2 - 60x + 450 = 0$

$=> 2(x^2 - 30x + 225) = 0$

$=> x^2 - 30x + 225 = 0$

$=> x^2 - 2 \times 15 \times x + (15)^2 = 0$

$=> (x - 15)^2 = 0$

$=> x = 15$

So, $x = 15$ and $y = 30 - x = 15$, and the area of $ABCD = 15 \times 15 = 225$.

The correct answer is Option B.

241. We now that the number of diagonals in a polygon with n sides = $\frac{n(n-3)}{2}$.

According to the problem statement, we have the number of diagonals equal to double the number of sides. Hence, we have the equation:

$\frac{n(n-3)}{2} = 2n$

$=> n(n - 3) = 4n => n(n - 3) - 4n = 0$

$=> n\{(n - 3) - 4\} = 0 => n(n - 7) = 0$

$=> n = 0$ or 7.

Clearly, $n \neq 0$.

Hence, $n = 7$.

The correct answer is option C.

242. Let the breadth of the rectangle be s inches.

We know that the difference between the length and breadth of the rectangle is 4 inches. This means that the length must be 4 inches longer than the breadth.

Hence, the length of the rectangle is $(s + 4)$ inches.

We know that the area of a rectangle is given by (length $\times$ breadth).

Thus, the initial area of the rectangle = $s(s + 4)$ square inches.

The increased length of the rectangle = $(s + 4) + 6 = (s + 10)$ inches.

Thus, the new area of the rectangle = $s(s + 10)$ square inches.

Hence, the increase in area = $s(s + 10) - s(s + 4) = s[(s + 10) - (s + 4)] = 6s$ square inches.

Hence, we have $6s = 120 => s = 20$.

Thus, the breadth of the rectangle is 20 inches.

The correct answer is Option B.

243. Let the initial length of a side of the square be s inches.

We know that the area of a square is given by the square of its side.

Thus, the initial area of the square = s^2 square inches.

The new length of a side of the square becomes $2s$ inches.

Thus, the new area of the square = $(2s)^2 = 4s^2$ square inches.

Hence, increase in area = $4s^2 - s^2 = 3s^2$ square inches.

Hence, we have $3s^2 = 300 => s^2 = 100 => s = 10$.

Thus, the initial side of the square is 10 inches.

The correct answer is Option A.

244. Assuming the radius of the cylinder to be r cm, its volume

$= \pi r^2 \times 84$

Thus, equating the volume of the sphere and the cylinder, we have:

$\frac{4}{3}\pi \times 7^3 = \pi r^2 \times 84$

$=> r^2 = \frac{4}{3} \times \frac{7^3}{84} = \frac{1}{3} \times \frac{7^3}{21} = \frac{7^2}{3^2}$

$=> r = \frac{7}{3}$ cm

(Taking the square root on both sides of the equation, the value $r = -\frac{7}{3}$ will be rejected because the radius of a physical object cannot be negative.)

The correct answer is choice B.

245. The perimeter of the triangle $= 2x + (x - 1) + (6 - x) = 2x + 5$

However, the value of x cannot be determined. Let us apply another approach to answer the question.

Since sum of any two sides is greater than the third side, we have:

- $2x + (x - 1) > 6 - x => x > 1.45$
- $(x - 1) + (6 - x) > 2x => x < 2.5$
- $(6 - x) + 2x > (x - 1) => 7 > 0$ - Always true (for all values of x)

Thus, we have:

$1.45 < x < 2.5$

Since x is an integer, the only possible value of $x = 2$

Thus, the sides are

$2x = 4$

$x - 1 = 1$

$6 - x = 4$

Thus, the perimeter of the triangle = 4 + 1 + 4 = 9

The correct answer is choice C.

5.4.2 Co-ordinate Geometry

246. For a line represented by $ax + by = k$, where the coefficient of y, i.e. b is POSITIVE, we have

- $ax + by \geq k$ represents the area ABOVE the line
- $ax + by \leq k$ represents the area BELOW the line

Let us analyze the lines:

- $2x - y \geq 0$: The coefficient of y is negative.
  Thus, modifying it, we have $y - 2x \leq 0$
  Thus, the required region is the one BELOW the line $y - 2x = 0$
- $x + y \leq 6$: The coefficient of y is positive.
  Thus, the required region is the one BELOW the line $x + y = 6$
- $y \geq 0$: The required region is above the X axis.

The diagram is shown below (the shaded area is the area of the region bounded by the lines $y = 2x$, $x + y = 6$ and the X axis.):

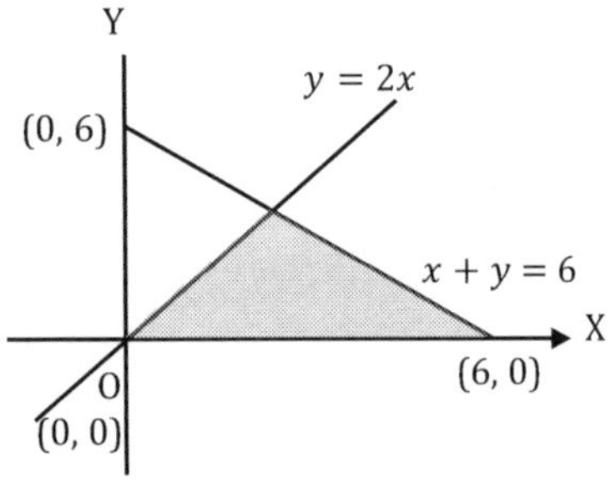

The correct answer is Option A.

Note: If required, we may also determine the area of the shaded region: Let us determine the point of intersection of these two lines:

Substituting $y = 2x$ in $x + y = 6$, we have

$x + 2x = 6 => x = 2$

$=> y = 2x = 4$

Thus, the coordinates of the point of intersection are $(2, 4)$.

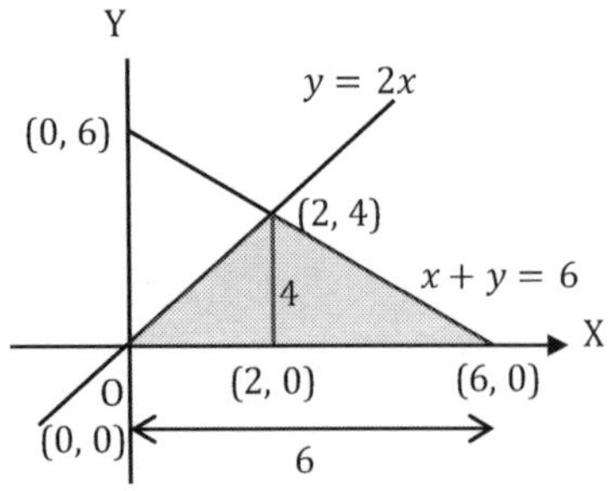

Thus, the required area = $\frac{1}{2} \times \text{Base} \times \text{Height} = \frac{1}{2} \times 6 \times 4 = 12$

Alternate approach:

Solving the equations $y = 2x$ and $x + y = 6$, we have $(x, y) = (2, 4)$.

Knowing this point of intersection, we pick a point from each option (shaded region), which would lie in that region and verify with each of the given inequalities:

Option A: A possible point lying in the shaded region would be (2, 3):

- $2x - y = 2 \times 2 - 3 = 1 > 0$ – Satisfies
- $x + y = 2 + 3 = 5 < 6$ – Satisfies
- $y = 3 > 0$ – Satisfies

Thus, option A is correct (Trying with the other shaded regions, one or more of the given inequalities would NOT be satisfied).

247. The equation of a circle having center at (p, q) and radius r is given by:

$$(x - p)^2 + (y - q)^2 = r^2$$

we have

$x^2 - 4x + y^2 + 6y = 12$

$\Rightarrow \{x^2 - 2(x)(2) + 2^2\} + \{y^2 + 2(y)(3) + 3^2\} - 2^2 - 3^2 = 12$

$\Rightarrow (x - 2)^2 + (y + 3)^2 = 12 + 4 + 9$

$\Rightarrow (x - 2)^2 + (y - (-3))^2 = 5^2$

Thus, the coordinates of the center of the circle are $(p, q) : (2, -3)$ and the radius of the circle $r = 5$.

Thus, we have

$p + q + r = 2 - 3 + 5 = 4$

The correct answer is 4.

248. Coordinates of the diagonal of the square = (3, 1) and (11, 7).

Thus, center of the circle is the mid-point of the diagonal, given by: $(\frac{3+11}{2}, \frac{1+7}{2}) = (7, 4)$.

Length of the diagonal = $\sqrt{(11 - 3)^2 + (7 - 1)^2} = \sqrt{8^2 + 6^2} = 10$

Thus, length of the side of the square = $\frac{10}{\sqrt{2}} = 5\sqrt{2}$

Thus, diameter of the in-circle = Length of the side of the square = $5\sqrt{2}$

Thus, radius of the circle = $\frac{5\sqrt{2}}{2} = \frac{5}{\sqrt{2}}$

Thus, the equation of the circle is:

$(x-7)^2+(y-4)^2=(\frac{5}{\sqrt{2}})^2$

$=> (x-7)^2+(y-4)^2=\frac{25}{2}$

$=> 2(x-7)^2+2(y-4)^2=25$

The correct answer is Option D.

249. We have

$\frac{x}{a}+\frac{y}{b}=1 => \frac{y}{b}=-\frac{x}{a}+1$

$=> y=\left(-\frac{b}{a}\right)x+b$

Thus, the slope of the above line is $(-\frac{b}{a})$

Also, we have:

$\frac{x}{p}+\frac{y}{q}=1 => \frac{y}{q}=-\frac{x}{p}+1$

$=> y=\left(-\frac{q}{p}\right)x+q$

Thus, the slope of the above line is $(-\frac{q}{p})$

Since the two lines are perpendicular to one another, the product of their slopes is '−1.'

$=> \left(-\frac{b}{a}\right)\times\left(-\frac{q}{p}\right)=-1$

$=> bq=-ap$

$=> ap+bq=0$

The correct answer is Option B.

250. Note: The equation $x^2+y^2=4=2^2$ represents a circle having center at origin (0, 0) with radius 2.

Since we need to determine the lattice points (points whose coordinates can take integral values) on the circle and inside it, we need to find the set of all points (x, y), such that:

$x^2+y^2\le 4$

This can be done in a tabular manner as shown below:

x	y	# of lattice points
0	$y^2\le 4 => y=0, \pm1, \pm2$	5
1	$y^2\le 3 => y=0, \pm1$	3
−1	$y^2\le 3 => y=0, \pm1$	3
2	$y^2\le 0 => y=0$	1
−2	$y^2\le 0 => y=0$	1
Total		**13**

The correct answer is 13.

251. Since AB is one side of the square, let us calculate its length.

AB = $\sqrt{(-3-0)^2 + (0-4)^2} = 5$

In a square with side a, the length of the diagonal is given by $a\sqrt{2}$.

Thus, the length of the diagonal of square ABCD = $5\sqrt{2}$

The correct answer is Option C.

252. Since the three points are collinear, i.e. lie on the same straight line, the lines formed by any two of the three points must have equal slope.

Thus, slope of the line joining $(1, 1)$ and $(a, 0)$ is the same as slope of the line joining $(1, 1)$ and $(0, b)$.

$=> \dfrac{1-0}{1-a} = \dfrac{1-b}{1-0}$

$=> \dfrac{1}{1-a} = \dfrac{1-b}{1}$

$=> 1 = 1 - a - b + ab$

$=> a + b = ab$

The correct answer is Option D.

253. The three vertices $(A, B \& O)$ are shown in the diagram below.

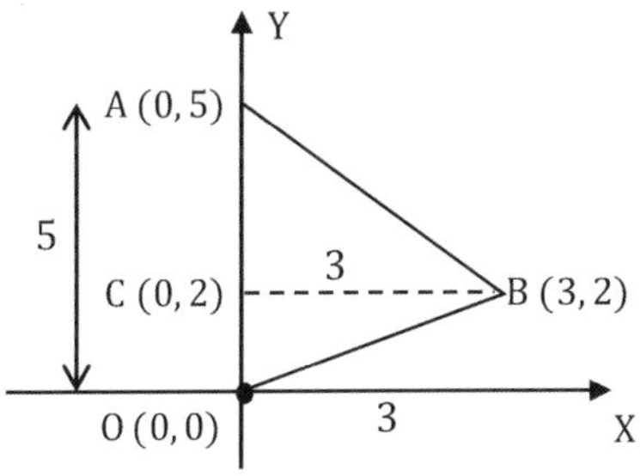

It can be seen above that both points A (0, 5) and O (0, 0) lie on the Y axis.

Taking AO to be the base of the triangle, BC is the height

Thus, area of the triangle

$= \dfrac{1}{2} \times \text{Base} \times \text{Height}$

$= \dfrac{1}{2} \times 5 \times 3$

$= 7.5$

The correct answer is 7.5.

254. Let us first calculate the lengths of the three sides of the triangle:

- $AB = \sqrt{(7-3)^2 + (9-(-7))^2} = \sqrt{16+256} = \sqrt{272}$
- $BC = \sqrt{(3-(-3))^2 + (-7-3)^2} = \sqrt{36+100} = \sqrt{136}$
- $CA = \sqrt{(-3-7)^2 + (3-9)^2} = \sqrt{100+36} = \sqrt{136}$

Thus, we see that: $AB^2 = BC^2 + CA^2$

Thus, Pythagoras' theorem is satisfied and the hypotenuse of the triangle is AB.

Thus, $\angle C$, i.e. the angle opposite the hypotenuse must be 90°.

Note: Had we noticed that: AB + BC = CA, then $\angle C = 0^\circ$; whereas, had it been AC + BC = AB, then $\angle C = 180^\circ$ (both situations are the degenerate case of a triangle becoming a straight line).

The correct answer is Option D.

255.

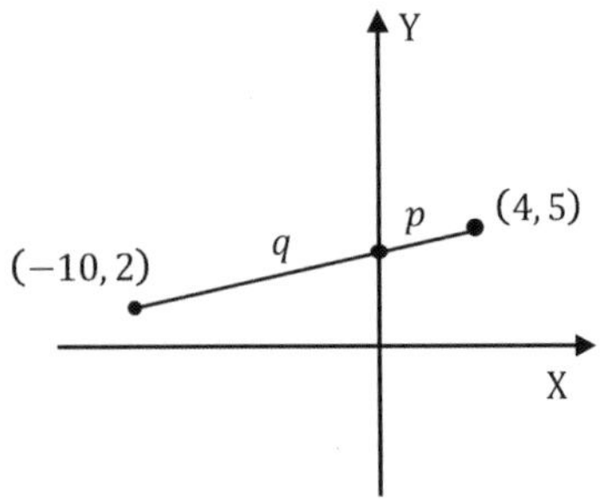

Let the Y axis intersect the line joining $(4, 5)$ and $(-10, 2)$ in the ratio $p : q$.

Thus,

Coordinates of the point of intersection $= \left(\dfrac{4p - 10q}{p+q}, \dfrac{5p + 2q}{p+q}\right)$

Since the intersecting point lies on the Y-axis, we have

$\dfrac{4p - 10q}{p+q} = 0 \Rightarrow 4p = 10q$

$\Rightarrow p : q = 5 : 2$

The correct answer is Option B.

5.4.3 Trigonometry

256. In this question, we need to calculate the length of the hypotenuse AC where the length of one side and an angle is given.

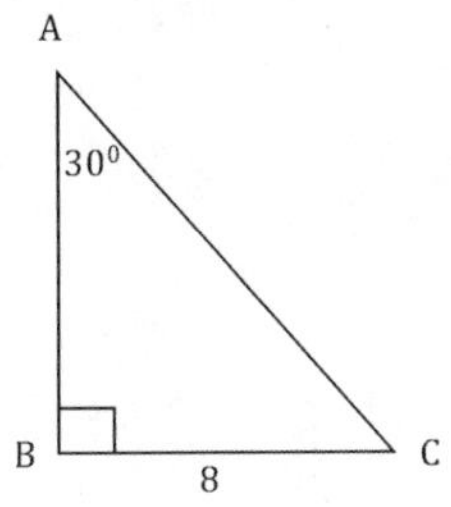

In order to find the hypotenuse, we need to use a trigonometric ratio that involves the hypotenuse. Both Sin θ and Cos θ are two such trigonometric ratios which involve the hypotenuse.

Now, we see that the given angle is angle A while the given side BC is opposite to it. Hence, we need to use Sin θ since it deals with the opposite side and the hypotenuse.

Sin 30° $= \frac{\text{opposite}}{\text{hypotenuse}} = \frac{\text{BC}}{\text{AC}}$.

Thus, we have $\frac{1}{2} = \frac{8}{\text{AC}}$ (since Sin 30° $= \frac{1}{2}$ & $BC = 8$)

$=> AC = 8 \times 2 = 16$.

Hence, the length of AC is 16.

The correct answer is Option C.

257. We know that in a right angled triangle, the square of the hypotenuse equals the sum of squares of the two legs of the triangle.

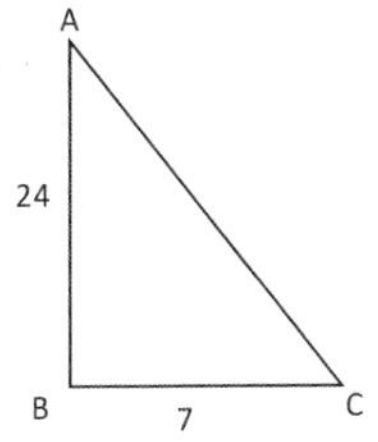

Thus, we have $AC^2 = AB^2 + BC^2$

$=> AC^2 = 24^2 + 7^2 = 576 + 49 = 625 = 25^2$

$=> AC = 25$.

The correct answer is Option C.

258. Let our triangle be ABC, right angled at B.

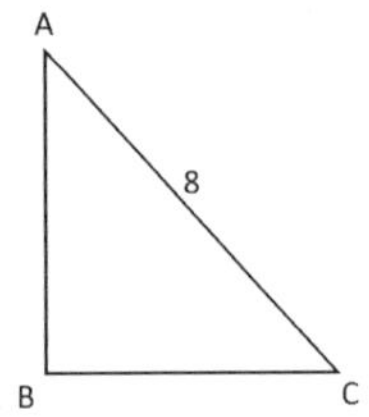

Since the triangle is isosceles, angles A and C must be equal.

Sum of the angles A and C = $180° - 90° = 90°$.

Hence, each angle A = C = $\frac{90°}{2} = 45°$.

In this question, we need to calculate the length of the AB and BC where the length of the hypotenuse is given.

In order to find the AB and AC, we need to use a trigonometric ratio that involves the hypotenuse. Both Sin θ and Cos θ are two such trigonometric ratios which involve the hypotenuse.

Thus, we have Sin A = Sin $45° = \frac{1}{\sqrt{2}} = \frac{BC}{AC} = \frac{BC}{8} \Rightarrow BC = \frac{8}{\sqrt{2}} = 4\sqrt{2}$.

Similarly, Sin C = Sin $45° = \frac{1}{\sqrt{2}} = \frac{AB}{AC} = \frac{AB}{8} \Rightarrow AB = \frac{8}{\sqrt{2}} = 4\sqrt{2}$.

Hence, sum of the other two sides = AB + BC = $4\sqrt{2}+4\sqrt{2} = 8\sqrt{2}$.

The correct answer is Option D.

259. To find AB, given the length of BC and angle A, we need to use tan θ since this trigonometric ratio relates the two legs of a right angled triangle.

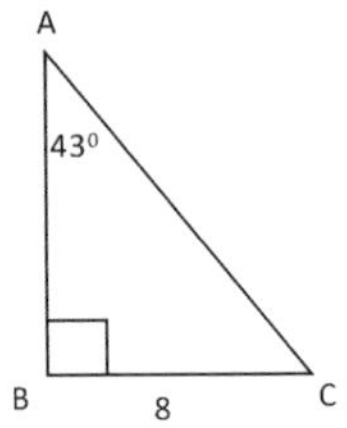

Thus, $\tan 43° = \frac{BC}{AB} \Rightarrow AB = \frac{BC}{\tan 43°} = \frac{8}{\tan 43°}$.

To find AC, given the length of BC and angle A, we need to use Sin θ since this trigonometric ratio relates the opposite side and the hypotenuse of a right angled triangle.

Thus, $\text{Sin } 43° = \frac{BC}{AC} \Rightarrow AC = \frac{BC}{\text{Sin } 43°} = \frac{8}{\text{Sin } 43°}$.

Thus, the difference between AB and AC = $AC - AB = \frac{8}{\text{Sin } 43°} - \frac{8}{\tan 43°}$.

The correct answer is Option A.

260.

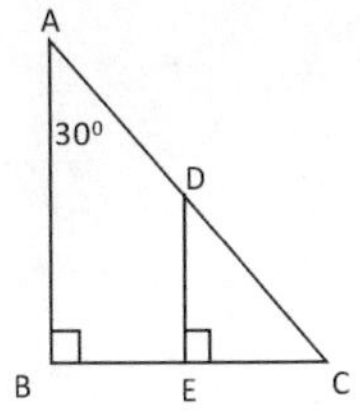

Since BE = EC, we have EC = $\frac{BC}{2} = \frac{8}{2} = 4$.

Also, angle C = 180° – (B + A) = 180° – (30°+90°) = 60°.

Now, in triangle DEC, Cos C = $\frac{EC}{DC}$ => Cos 60° = $\frac{4}{DC}$ => $\frac{1}{2} = \frac{4}{DC}$ => DC = 8.

The correct answer is Option B.

261. In the triangle, corresponding to angle A, we know the values of the opposite side and the hypotenuse. Hence, we work with Sin A.

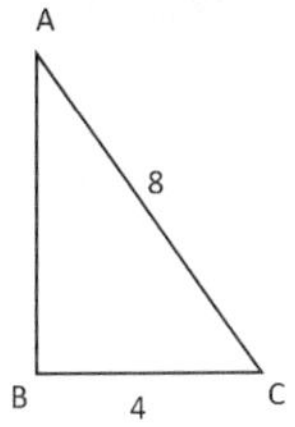

Sin A = $\frac{BC}{AC} = \frac{4}{8} = \frac{1}{2}$.

We know that Sin 30° = $\frac{1}{2}$.

Thus, A = 30°.

The correct answer is Option A.

262.

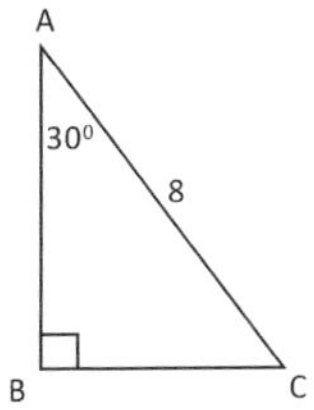

Let the garden be the triangle ABC as shown beside, right angled at B.

For a right angled triangles triangle, the longest side is the hypotenuse = 8 m long.

We need to use Sin θ and Cos θ to find the other two sides of the triangle since only these two ratios relate the sides to the hypotenuse.

Sin 30° = $\frac{BC}{AC}$ => $\frac{1}{2} = \frac{BC}{8}$ => BC = 4 m.

Cos 30° = $\frac{AB}{AC}$ => $\frac{\sqrt{3}}{2} = \frac{AB}{8}$ => AB = $4\sqrt{3}$ m.

Thus, perimeter of the triangle = AB + BC + AC = $4\sqrt{3}$+4 + 8 = $(12 + 4\sqrt{3})$ m.

The correct answer is Option D.

263.

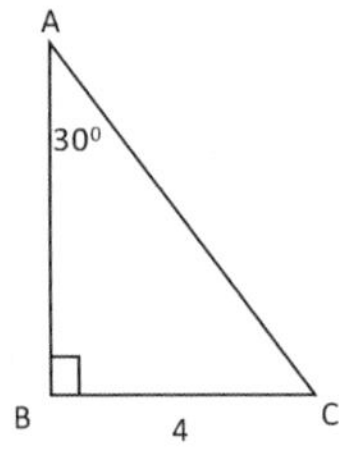

Let the triangle be ABC, right angled at B.

Let angle A be 30°.

Thus, the other angle C must be 180° − (90° + 30°) = 60°.

The smallest side must be opposite to the smallest angle.

Hence, we have BC = 4 (the smallest side).

We know that area of a triangle is gives as: Area = $\frac{1}{2}$×Base × Height.

Here, we can take BC as the base and AB as the height.

Thus, we need to find the length of AB.

Since we need the relation between the two legs of the right angled triangle, we work with tan θ.

Thus, tan A = tan 30° = $\frac{BC}{AB}$ => $\frac{1}{\sqrt{3}} = \frac{4}{AB}$ => AB = $4\sqrt{3}$.

Hence, area of the triangle = $\frac{1}{2}$×AB × BC = $\frac{1}{2}$×$4\sqrt{3}$ × 4 = $8\sqrt{3}$.

The correct answer is Option B.

264.

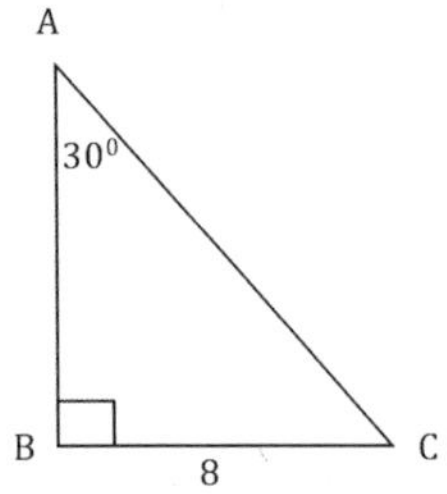

Here, we need to focus on triangle BDC in order to find the length of BD.

We know that angle B is 90° and angle A is 30°.

Thus, angle C = 180° − (90° + 30°) = 60°.

In triangle BDC, with respect to angle C, BD is the opposite side and BC is the hypotenuse.

Thus, in order to relate these two sides, we use Sin θ.

Thus, Sin C = Sin 60°= $\frac{BD}{BC}$ => $\frac{\sqrt{3}}{2} = \frac{BD}{4}$ => BD = $2\sqrt{3}$.

The correct answer is Option C.

265. Let the balloon be at the point A as shown in the diagram below.

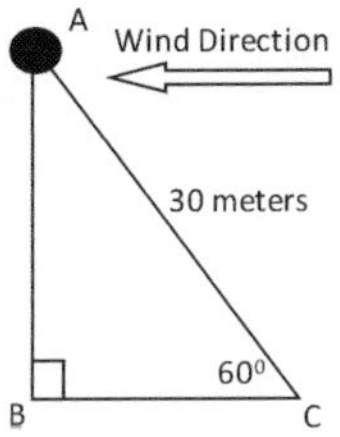

AC is the rope with which the balloon is tied to the ground (the ground being BC).

Due to the wind, the balloon sways and its rope makes an angle of 60° with the ground.

We need to calculate the current height of the balloon above the ground level i.e. AB.

Thus, in triangle ABC, with reference to angle C, we know the hypotenuse and we need to calculate the opposite side. Hence, we work with Sin θ.

Thus, we have Sin 60° = $\frac{AB}{AC}$=> $\frac{\sqrt{3}}{2} = \frac{AB}{30}$=> AB = $15\sqrt{3}$ meters.

The correct answer is Option D.

266. We can simplify the above as follows:

$$\frac{3\pi}{4}^{c} + \frac{11\pi}{5}^{c} - \frac{7\pi}{10}^{c} = \left(\frac{3\pi \times 5 + 11\pi \times 4 - 7\pi \times 2}{20}\right)^{c} = \left(\frac{15\pi + 44\pi - 14\pi}{20}\right)^{c} = \left(\frac{45\pi}{20}\right)^{c} = \left(\frac{9\pi}{4}\right)^{c}$$

We know that π radian is equivalent to 180°.

Thus, $\frac{9\pi}{4}$ radian is equivalent to $180° \times \frac{9}{4} = 405°$.

Thus, the required value is 405°.

The correct answer is Option C.

267. We can simplify the above as follows:

$$\frac{7\pi}{4}^{c} + \frac{4\pi}{3}^{c} - \frac{5\pi}{12}^{c} - 180° = \left(\frac{7\pi \times 3 + 4\pi \times 4 - 5\pi \times 1}{12}\right)^{c} - 180°$$

$$=> \left(\frac{21\pi + 16\pi - 5\pi}{12}\right)^{c} - 180° = \left(\frac{32\pi}{12}\right)^{c} - 180° = \frac{8\pi}{3}^{c} - 180°$$

We know that π radian is equivalent to 180°.

Thus, $\frac{8\pi}{3}$ radian is equivalent to $180° \times \frac{8}{3} = 480°$.

Thus, $\frac{8\pi}{3}^{c} - 180° = 480° - 180° = 300°$. (We do not find an option with 300°)

Converting back to radians: Since $180^\circ = \pi$ radians, we have:

$300^\circ = \left(\frac{\pi}{180} \times 300\right)^c = \frac{5\pi}{3}^c$.

Thus, the required value is 300° or $\frac{5\pi}{3}^c$.

The correct answer is Option B.

268. We know that π radians are equivalent to 180°.

Thus, $\frac{11\pi}{3}$ radians are equivalent to $180^\circ \times \frac{11}{3} = 660^\circ$.

Hence, the first statement is correct.

The unit circle along with the quadrants marked is shown below. The values of the angles at the four major points are also shown.

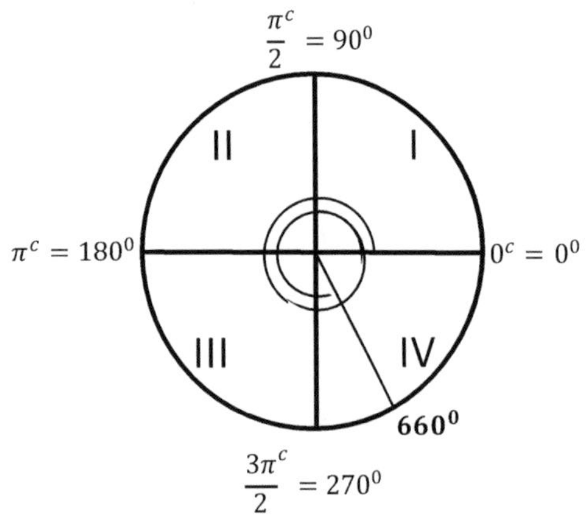

Now, $660^\circ = 360^\circ \times 1 + 300^\circ$.

Thus, in order to trace the angle, we need to make one full revolution from 0° and then cover 300° extra. In the diagram, the angle corresponding to 660° => 300° is shown.

Thus, we finally end in the fourth quadrant. Thus, the second statement is not correct.

We also know that in the fourth quadrant, the value of Cos θ is positive.

Thus, the third statement is correct.

The correct answer is Option D.

269. We know that π radians are equivalent to 180°.

Thus, $\frac{13\pi}{4}$ radians are equivalent to $180^\circ \times \frac{13}{4} = 585^\circ$.

Hence, the first statement is correct.

The unit circle along with the quadrants marked is shown below. The values of the angles at the four major points are also shown.

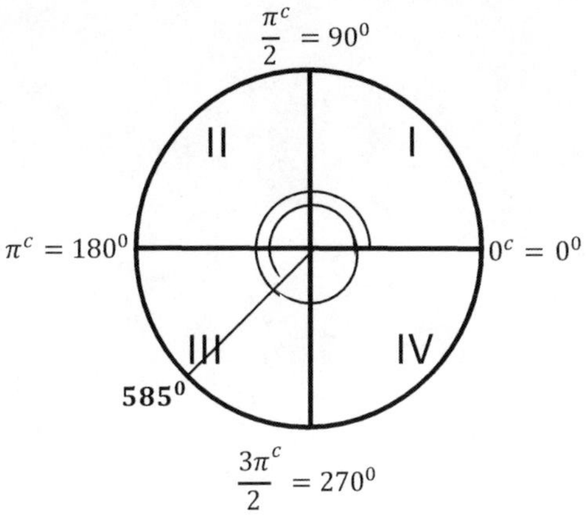

Now, $585^\circ = 360^\circ \times 1 + 225^\circ$.

Thus, in order to trace the angle, we need to make one full revolution from 0° and then cover 225° extra. In the diagram, the angle corresponding to $585^\circ \Rightarrow 225^\circ$ is shown.

Thus, we finally end in the third quadrant.

We know that in the third quadrant, the value of $\tan\theta$ is positive while the value of $\text{Cos}\theta$ is negative.

Thus, the second and third statements are not correct.

The correct answer is option A.

270. We can see that the value of Cos θ is positive.

We know that the value of Cos θ is positive in the first and fourth quadrants only. Thus, θ should lie in either the first or fourth quadrants.

Hence, the first statement is not correct.

Again, the value of tan θ is positive in the first quadrant and is negative in the fourth quadrant. Since θ lies in either the first or fourth quadrant, the value of tan θ can be both positive as well as negative.

Hence, the second statement is correct.

Again, the value of Sin θ is positive in the first quadrant and negative in the fourth quadrant. Since θ lies in either the first or fourth quadrant, the value of Sin θ may be both positive as well as negative.

Hence, the third statement is not correct.

The correct answer is Option B.

271. We know that π radians are equivalent to 180°.

Thus, $\frac{7\pi}{6}$ radians are equivalent to $180^\circ \times \frac{7}{6} = 210^\circ$.

The above unit circle along with the quadrants marked is shown below. The values of the angles at the four major points are also shown.

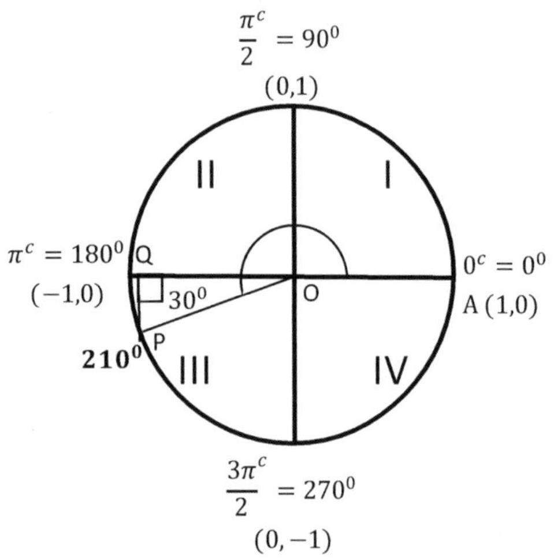

In the diagram, the angle corresponding to 210° is also shown.

Thus, we finally end in the third quadrant.

We know that in the third quadrant, the values of both $X$and Y coordinates are negative.

Thus, Option D, which has both coordinates as negative, must be the correct answer.

The correct answer is Option D.

Calculation

The angle the point makes with the negative X-axis is 30°.

Thus, Sin $30^\circ = \frac{PQ}{OP} \Rightarrow \frac{1}{2} = \frac{PQ}{1} \Rightarrow PQ = \frac{1}{2}$

=> Y-coordinate of P is $(-1/2)$

Cos $30^\circ = \frac{OQ}{OP} \Rightarrow \frac{\sqrt{3}}{2} = \frac{OQ}{1} \Rightarrow OQ = \frac{\sqrt{3}}{2}$

=> X-coordinate of P is $-\frac{\sqrt{3}}{2}$.

272. We know that π radians are equivalent to 180°.

Thus, $\frac{5\pi}{4}$ radians are equivalent to $180^\circ \times \frac{5}{4} = 225^\circ$.

The above unit circle along with the quadrants marked is shown below. The values of the angles at the four major points are also shown.

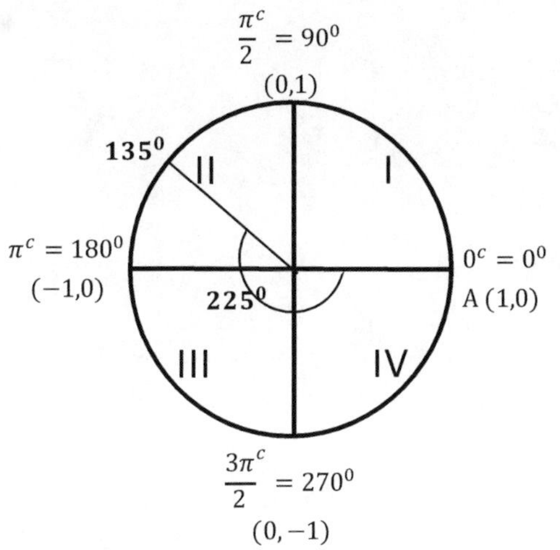

In the diagram, the angle corresponding to 225° in the clockwise direction is shown.

One should note that 225° in the clockwise direction is equivalent to $360° - 225° = 135°$.

Thus, we finally end in the second quadrant.

We know that in the second quadrant, the value of X coordinate is negative while that of the Y coordinate is positive.

Thus, Option B, which satisfies the above conditions, must be the answer.

The correct answer is Option B.

273. We know that π radians are equivalent to 180°.

Thus, $\frac{4\pi}{5}$ radians are equivalent to $180° \times \frac{4}{5} = 144°$ and $\frac{5\pi}{6}$ radians are equivalent to $180° \times \frac{5}{6} = 150°$.

Thus, we have $144° \leq \theta \leq 150°$ i.e. θ lies in the second quadrant.

The angle is shown in the diagram below.

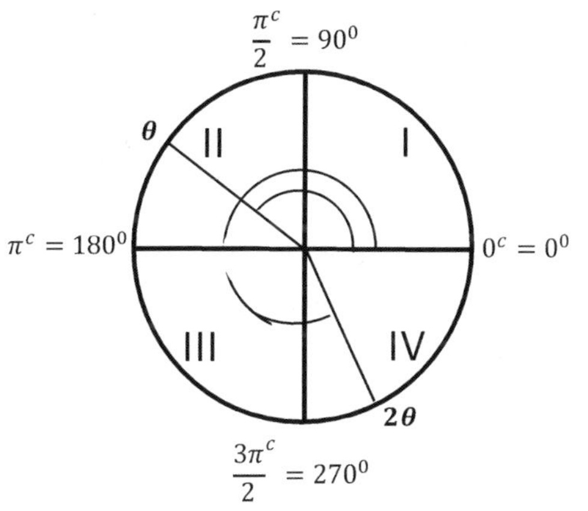

Thus, the value of Cos θ is negative.

Hence, the first statement is not correct.

Again, since $144^\circ \leq \theta \leq 150^\circ$, we have $2 \times 144^\circ \leq 2\theta \leq 2 \times 150^\circ$ => $288^\circ \leq 2\theta \leq 300^\circ$.

Thus, 2θ lies in the fourth quadrant as shown in the diagram.

Thus, the value of Cos (2θ) is positive and the value of Sin (2θ) is negative.

Hence, both second and third statements are correct.

The correct answer is Option C.

274. We know that π radians are equivalent to 180°.

Thus, $\frac{3\pi}{4}$ radians are equivalent to $180^\circ \times \frac{3}{4} = 135^\circ$ and $\frac{4\pi}{5}$ radians are equivalent to $180^\circ \times \frac{4}{5} = 144^\circ$.

Thus, $135^\circ < \theta < 144^\circ$ i.e. θ lies in the second quadrant.

The angle is shown in the diagram below.

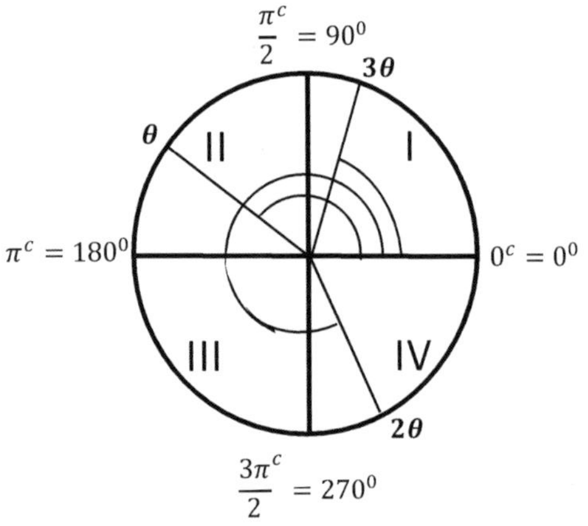

Thus, the value of tan θ is negative.

Hence, the first statement is correct.

Again, since $135^\circ < \theta < 144^\circ$, we have $2 \times 135^\circ < 2\theta < 2 \times 144^\circ$ => $270^\circ < 2\theta < 288^\circ$.

Thus, 2θ lies in the fourth quadrant as shown in the diagram above.

Thus, the value of tan (2θ) is negative.

Hence, the second statement is correct.

Again, since $135^\circ < \theta < 144^\circ$, we have $3 \times 135^\circ < 3\theta < 3 \times 144^\circ$ => $405^\circ < 3\theta < 432^\circ$.

Since the angles exceed 360°, we remove 360° to get:

$405^\circ - 360^\circ < 3\theta < 432^\circ - 360^\circ$

=> $45^\circ < 3\theta < 72^\circ$

Thus, 3θ lies in the first quadrant as shown in the diagram above.

Thus, the value of tan (3θ) is positive.

Hence, the third statement is correct.

Thus, all statements are correct.

The correct answer is Option D.

275. We have $\theta = 150° => 2\theta = 300°$, $4\theta = 600°$ and $5\theta = 750°$.

Let us first plot the angles θ, 2θ, 4θ and 5θ on the unit circle.

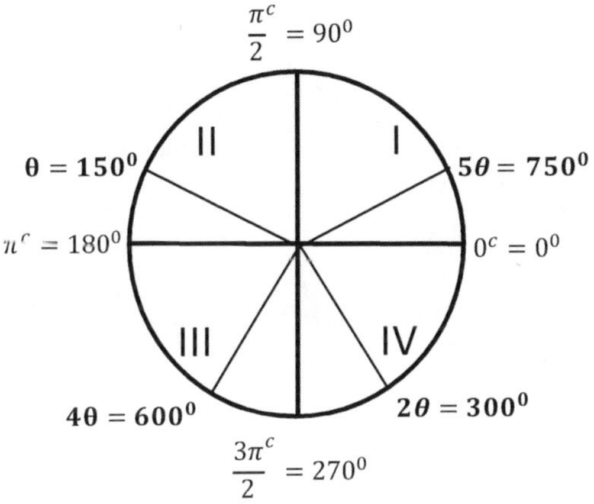

Thus, we can see that θ lies in the second quadrant, 2θ lies in the fourth quadrant, 4θ lies in the third quadrant and 5θ lies in the first quadrant.

Verifying each statement:

Statement I: Sin (2θ) is negative (angle in fourth quadrant) and Cos (4θ) is negative (angle in third quadrant).

Thus, both have the same sign.

Hence, the first statement is correct.

Statement II: tan (2θ) is negative (angle in fourth quadrant) and Cos (5θ) is positive (angle in first quadrant).

Thus, they have opposite signs.

Hence, the second statement is correct.

Statement III: Sin (4θ) is negative (angle in third quadrant) and tan (5θ) is positive (angle in first quadrant).

Thus, they have opposite signs.

Hence, the third statement is not correct.

The correct answer is Option C.

Chapter 6

Talk to Us

Have a Question?

Email your questions to *info@manhattanreview.com*. We will be happy to answer you. Your questions can be related to a concept, an application of a concept, an explanation of a question, a suggestion for an alternate approach, or anything else you wish to ask regarding the DSAT.

Please do mention the page number when quoting from the book.

Best of luck!

Professor Dr. Joern Meissner
& The Manhattan Review Team

Made in United States
Orlando, FL
04 November 2026

90202251R00157